HOT

BOOKS BY MARK HERTSGAARD

*Nuclear, Inc.: The Men and Money
Behind Nuclear Energy*

*On Bended Knee: The Press and
the Reagan Presidency*

*A Day in the Life: The Music and
Artistry of the Beatles*

*Earth Odyssey: Around the World in Search
of Our Environmental Future*

*The Eagle's Shadow: Why America Fascinates
and Infuriates the World*

*Hot: Living Through the Next
Fifty Years on Earth*

HOT

Living Through the

Mark Hertsgaard

Next Fifty Years on Earth

HOUGHTON MIFFLIN HARCOURT

BOSTON   NEW YORK   2011

For information about permission to reproduce selections from this book,
write to Permissions, Houghton Mifflin Harcourt Publishing Company,
215 Park Avenue South, New York, New York 10003.

www.hmhbooks.com

Library of Congress Cataloging-in-Publication Data
Hertsgaard, Mark, date.
Hot : living through the next fifty years on earth / Mark Hertsgaard.
p. cm.
Includes index.
ISBN 978-0-618-82612-4
1. Global warming. 2. Climatic changes. 3. Global environmental change.
4. Human beings—Effect of climate on. I. Title.
QC981.8.G56H47 2010
304.2'5—dc22   2010012416

Book design by Brian Moore

Printed in the United States of America

DOC 10 9 8 7 6 5 4 3 2 1

DEDICATION

For my daughter, Chiara, who has to live through this

Contents

HOT

Prologue: Growing Up Under Global Warming

Working on climate change used to be about saving the world for future generations. Not anymore. Now it's not only your daughter who is at risk, it's probably you as well.

— MARTIN PARRY, co-chair of the *Fourth Assessment Report*,
Intergovernmental Panel on Climate Change

I covered the environmental beat for fifteen years before I became a father. Much of that time was spent overseas, where, like many other journalists, I saw more than my share of heartbreaking things happening to children. But they were always other people's children.

My first time was in the old Soviet Union, where I exposed a series of nuclear disasters that had been kept secret for decades by both the KGB and the CIA. One day, I visited the leukemia ward of the local children's hospital, where a dozen mothers and children had gathered to speak with me. Many of the kids were bald, thanks to the chemotherapy that was now being applied in a last-gasp attempt to save their stricken bodies. The mother of one heavyset girl could not stop sobbing. When her daughter stroked her arm to comfort her, the mother unleashed a

deep, aching wail and fled the room. This woman, like the other mothers, knew what the children did not: the doctors expected 75 percent of
these children to be dead within five years.

Soon after, I spent four months in the northeastern Horn of Africa,
mainly covering drought and civil war. It was there, in a refugee camp in
southern Sudan in 1992, that I first came face-to-face with starving children. In my mind's eye, I can still see the young mother as she entered
the Red Cross compound, hoping to see a nurse. Unfolding the tattered
cloth she had slung from her neck, the mother revealed a nine-month-
old baby girl, a tiny creature with a grotesquely large skull and legs no
thicker than my fingers. Like one of every eleven African children, this
poor child would not live to see her first birthday.

Later still I visited China, where millions of children were breathing and drinking some of the most carcinogenic air and water on the
planet. Crisscrossing the country in 1996 and 1997, I became the first
writer to describe China's emergence as a climate change superpower,
second only to the United States. To fuel its explosive economic growth
and lift its people out of poverty, China was burning more coal than any
other nation on earth, making its skies toxic and dark even on sunny
afternoons. Some of the worst health effects were being measured in the
northern industrial city of Shenyang. One afternoon I visited a heavy-
machinery factory that ranked among the city's deadliest polluters. I arrived just in time to see the street fill with hundreds of children. Chat-
tering and laughing, they walked in rows six abreast, returning home
from school, inhaling poison with every breath.

In my journalism, I tried to draw the outside world's attention to the
plight of all of these children, as well as to its causes and potential remedies. Emotionally, though, I could keep a distance. This was partly because, as I say, these were other people's children. But it was also, I now
see, because I was not yet a parent myself. I did not really understand,
viscerally, how it feels to see one's own child be sick, in danger, and perhaps facing death.

I found out soon enough.

My daughter was born in 2005, in San Francisco, at the end of a long
and difficult labor. After many hours and much pushing and tugging,
she finally emerged from her mother's body. By that time, the urgency

of the situation had drawn a dozen nurses into the room. As they attended to their various tasks—lifting the baby onto her mother's chest, administering her first bath—one nurse after another made the same observation.

"Wow, look how alert this baby is," the nurse in charge commented.

"I know," marveled a colleague. "Look at her eyes!"

Apparently, most newborns keep their eyes shut against the light of the new world. Not ours. Her blazing blue eyes were wide open. From the moment she got here, this little girl was awake on the planet.

When it came time to give her a name, her mother and I remembered these first moments of her life and decided to call her Chiara. In the Italian language of her ancestors, Chiara (pronounced with a hard C, Key-AR-a) means "clear and bright."

Everything seemed fine until two days later. We had taken Chiara home from the hospital. As scheduled, a nurse came to conduct a follow-up exam. A few hours later, a doctor called and told us to bring Chiara back to the hospital, to the intensive care unit, right away. The exam had found dangerous levels of bilirubin in her blood. Brain damage or worse could follow.

At the intensive care unit, Chiara was placed inside an incubator, a white gauze headband stretched around her little skull to protect her eyes. The nurses jokingly called it a raccoon mask. Day and night I sat beside the incubator, watching Chiara's yellowish body get drenched with vitamin D–laden light.

Yet as worried as I was, I also felt fortunate. Unlike the children I recalled in Russia, Africa, and China, Chiara had access to excellent medical care. Within three days, she had completely recovered, with no lasting damage, and was sent back home.

Six months later, though, a different threat arose to my daughter's life, and this time no quick fix was available. During a reporting trip to London in October 2005, I learned that the global warming problem had undergone a momentous transformation. Humanity, it turned out, was in a very different fight than most people realized. Now, no matter what we did, Chiara and her generation were fated to inherit—indeed, spend most of their lives coping with—a climate that would be hotter than ever before in our civilization's history.

Global Warming Triggers Climate Change

The most important interview I did in London was with Sir David King, the chief science adviser to the British government. King received me at his office high above Victoria Street, a few blocks west of Parliament. When he stood up to shake hands, I could glimpse the spires of Westminster Abbey over his shoulder. Though not a tall man, King projected an unmistakable air of command as he invited me to join him at a conference table. I was on assignment from *Vanity Fair* magazine, a fact that seemed to amuse King, who had chaired the chemistry department at Cambridge University for seven years before entering government. "That's one publication I never thought I'd appear in," he said, chuckling. "I guess climate change has finally made the mainstream in the United States."

Since becoming science adviser in 2000, David King had done as much to raise awareness of climate change as anyone except former U.S. vice president Al Gore. Among other accomplishments, King had reportedly persuaded Prime Minister Tony Blair to make the issue a priority, and Blair in turn made climate change the lead topic at the 2005 summit of the Group of Eight, the world's richest economies. King also had a gift for attracting media coverage. In 2004, he called climate change "the most severe problem we are facing today—more serious even than the threat of terrorism." Coming barely two years after the September 11 terrorist attacks, the comment enraged right-wingers in Washington. But King told me he "absolutely" stood by it. "I think this is a massive test for our civilization," he said. "Our civilization has developed over the past eight thousand years during a period which has had remarkably constant weather conditions and remarkably constant ocean levels. What is happening now, through our use of fossil fuels, through our growing population, is that that stable period is under severe threat."

I had begun following the climate issue in 1989, the year I first interviewed James Hansen. As the chief climate scientist at the space agency NASA, Hansen had put climate change on the international agenda the year before when, in testimony to the U.S. Senate, he declared that man-made global warming had begun. Of course, *natural* global warming had been taking place for a very long time already. Building on the work of scientists going back to Joseph Fourier in 1824, the Nobel Prize–winning chemist Svante Arrhenius had published a theory of the greenhouse effect in 1896. The theory held that carbon dioxide and other gases in the atmosphere trap heat from the sun that otherwise would escape back into space, thus raising temperatures on earth. Indeed, without the greenhouse effect, Earth would be too cold to support human life. In his Senate testimony, Hansen argued that human activities—notably, the burning of oil, coal, and other carbon-based fuels—had now added excessive amounts of carbon dioxide to the atmosphere. This extra CO_2 was raising global temperatures, and they would rise significantly higher if emissions were not reduced. The higher temperatures in turn could trigger dangerous climate change, Hansen added.

A quick word here on definitions: although the terms *global warming* and *climate change* are often used interchangeably, a critical difference exists between them. In this book, *global warming* refers to the man-made rise in temperatures caused by excessive amounts of carbon dioxide, methane, and other greenhouse gases in the atmosphere. *Climate change*, on the other hand, refers to the effects these higher temperatures have on the earth's natural systems and the impacts that can result: stronger storms, deeper droughts, shifting seasons, sea level rise, and much else. To oversimplify slightly, think of global warming as the equivalent of a fever and climate change as the aches, chills, and vomiting the fever can cause.

It was partly Hansen's 1988 Senate testimony that led me to spend most of the 1990s traveling around the world, researching humanity's environmental future. I was also motivated by interviews I had done with Jimmy Carter, the former U.S. president; Jacques Cousteau, the French underwater explorer; Lester Brown, the founder of the Worldwatch Institute; and other leading environmental thinkers. Brown in particular had argued that problems such as global warming and population growth were cumulative in nature and thus presented a new kind of environmental challenge: if they were not reversed within the next ten years, Brown said, they could acquire too much momentum to reverse at all. I wasn't necessarily convinced Brown was correct, but his assertion was a provocative hypothesis to explore as I set off around the world. My mission was to investigate whether our civilization's survival was indeed threatened by global warming, population growth, and related environmental hazards. And if the danger was real, I hoped to gauge whether human societies would act quickly and decisively enough to avoid environmental self-destruction.

Over the course of six years, I investigated conditions at ground level in sixteen countries in Asia, Africa, Europe, and North and South America to write the book *Earth Odyssey*. As part of my research, in 1992 I covered the UN "Earth Summit" in Brazil, where I watched the heads of state or government for most of the world's nations (including the United States, under the first President Bush) affix their signatures to the UN Framework Convention on Climate Change. This treaty remains in force today; the better-known Kyoto Protocol is an amendment to it. The treaty's key sentence affirmed the world's governments' pledge to keep atmospheric levels of greenhouse gases low enough to "prevent dangerous anthropogenic [man-made] interference with the climate system."

From the start, then, the goal of the international community was to stop global warming *before* it triggered dangerous climate change. As the 1990s wore on, more and more scientists came to agree with Hansen that average global temperatures were rising and that humanity's greenhouse gas emissions were the main reason why. But—and this is the key point—most scientists did not expect this global warming to trigger significant climate change for a long time to come: the year 2100 was the date usually referenced in scientists' studies of sea level rise, famine, and other possible impacts. Although 2100 was chosen partly because it was distant enough to enable more reliable computer modeling studies, the date had the practical effect of implying—especially to politicians, journalists, ordinary citizens, and non-scientists in general—that serious impacts were a century away. In short, climate change was regarded as a grave but remote future threat, and one that could still be averted if humanity reduced emissions in time.

Meanwhile, a tiny but well-funded minority had begun arguing that global warming was little more than a politically inspired hoax. Frederick Seitz, a former president of the U.S. National Academy of Sciences, was the highest-ranking scientist making this claim, but most of the argument was carried by spokespersons for the Global Climate Coalition, a pressure group created and funded by U.S.-based energy and auto companies. Notwithstanding its studiously neutral name, the coalition would spend millions of dollars in the 1990s on a public disinformation campaign whose strategy and tactics recalled the tobacco industry's earlier efforts to persuade people that smoking cigarettes does not cause cancer. Indeed, Seitz and organizations he directed were paid more than $45 million for their work, first by tobacco and later by energy companies, as I'll describe later in this book.

The goal of the disinformation campaign was to "reposition global warming as theory rather than fact," according to an internal strategy memo unearthed by journalist Ross Gelbspan, who exposed the campaign in his 1997 book *The Heat Is On*. Despite such revelations, the deniers had considerable influence over the public debate, at least in the United States. Fortified by corporate contributions and bipartisan support in the U.S. Congress, deniers turned global warming into a political rather than a scientific dispute, blaming a supposed conspiracy by Gore and other "liberals" to advance a radical environmental agenda. James Inhofe, a Republican senator from the oil-rich state of Oklahoma, led the charge, calling global warming "the greatest hoax ever perpetrated on the American people." But Inhofe, Seitz, and other deniers could never have fooled the public and stalled political progress without the help of the mainstream media. In the name of providing journalistic balance, U.S. news stories routinely gave as much prominence to deniers of man-made global warming as they did to affirmers of it, even though the deniers amounted to a tiny fraction of the scientific community and often, as in Seitz's case, were in the pay of fossil fuel companies.

The upshot was that public discussion of global warming from the 1990s onward was framed as an if-then formulation: *if* global warming is real, and *if* greenhouse gas emissions are not reduced, then humanity *might* face problems in the far-off future.

In our London interview, David King shattered this framing. Climate change, the science adviser told me, was no longer a distant hypothetical threat: it had already begun. What's more, climate change was guaranteed to get worse, perhaps a lot worse, before it got better.

No comparably prominent scientist in the United States was saying this sort of thing publicly in 2005. In particular, King's assertions went beyond the findings of the Intergovernmental Panel on Climate Change (IPCC), an international group of scientists and experts the UN had created in 1988 to advise the world's governments on global warming. The IPCC had issued three major reports on climate change by the time I interviewed King. Its *First Assessment Report* appeared in 1990, its *Second Assessment Report* in 1995, and its *Third Assessment Report* in 2001. Only in its *Fourth Assessment Report,* released in 2007, eighteen months after our interview, did the IPCC declare that the scientific evidence for man-made global warming was "unequivocal" and that long-term sea level rise and other impacts of climate change had become inevitable. If King was ahead of the curve, it was partly because, as the British government's chief science adviser, he kept a close eye on what his country's scientists were doing. Indeed, he told me, a group of British scientists had recently detected the so-called climate signal; that is, the scientists had demonstrated that global warming had already exerted an impact on the earth's climate that stood out from the statistical noise of the historical record.

The scientific rule of thumb had always been that no single weather event could be linked to global warming. After all, extreme weather events were a recurring fact of history; how could one know whether a given event was caused by global warming, not by something else? But Britain, King claimed, had some of the best weather scientists in the world, a legacy of the nation's past as a maritime empire. Now, three of those scientists—Peter A. Stott, D. A. Stone, and M. R. Allen—had produced a breakthrough of epochal significance.

Their research, published in the scientific journal *Nature,* focused on the summer of 2003, when Europe experienced a brutal heat wave. Public health systems were overwhelmed. By mid-August, corpses were piling up outside morgues in Paris. The summer of 2003 was not only "the hottest . . . on record," King told me, it was also "the deadliest disaster in modern European history." It left 31,000 people dead, he added—a death toll twenty times higher than that of Hurricane Katrina, which had struck six weeks before our interview.

"Now," said King, "if we treat that hot summer as a single extreme event, the conclusion is that it's a 1-in-800-year event—quite a highly unlikely event." But the science adviser pointed out that over the past fifty years global warming had created "a rising baseline" of higher temperatures, which was heating up Europe's weather in both normal and abnormal years. Thus Europe's average summer temperature in 2005 was the same as it had been in the hottest summer of the twentieth century, in 1947. When an extreme event like the 2003 summer came along, the rising baseline made it even hotter. The conclusion, said King, was that "about half" of the excessive heat Europeans endured in 2003 was due to the rising baseline— that is, to global warming.

King then recalled the single harshest truth about climate change: we can't turn it off, at least not anytime soon. Once global warming has triggered it, climate change continues for a very long time. The reason? The laws of physics and chemistry—what King called "the inertia of the climate system." Carbon dioxide, the most plentiful greenhouse gas, stays in the atmosphere for as long as hundreds of years; oceans absorb the heat created by global warming and release it back to the atmosphere over the course of centuries. As a result, there is a lag effect, a delay, between the time greenhouse gas emissions may be reduced and the time global temperatures may begin to fall. The lag effect meant that Europe was already locked in to more frequent heat waves in the years ahead. Because of the rising baseline of temperatures, the science adviser told me, by 2050 Europe was projected to experience heat waves like that of 2003 once every two years.

King summarized the dilemma by offering a hypothetical case: even if our civilization stopped emitting all carbon dioxide overnight, he said, "temperatures will keep rising and all the impacts [storms, drought, sea level rise, and so on] will keep changing for about twenty-five years."

I asked if that meant it was "too late to save humanity, if that's not being too dramatic."

"No, no, it's not too late," King hurried to reply. "And saving humanity, I think, is not being too dramatic." Because we had waited so long to reduce emissions, we now had no choice but "to adapt to the impacts that are in the pipeline," King said. At the same time, the longer we wait to cut emissions, the greater the impacts will be. "So let's never give up on this," he said.

I had a six-month-old daughter, I replied, so giving up was not an option.

"Right," he said, flashing a quick smile. "My kids know who I'm battling for."

"Chiara Has to Live Through This"

After leaving King's office, I needed time to absorb all I had heard, so I headed down Victoria Street to walk along the river Thames. The weather was sunny, pleasant, a stark contrast to King's dire pronouncements. In effect, the science adviser had told me that climate change had already arrived, a hundred years ahead of schedule. If he was right, the debate over global warming was forever altered.

If climate change had indeed already begun, the inertia of the climate system ensured that the planet was locked in to at least twenty-five more years of rising temperatures no matter what—no matter how many solar panels people bought, no matter how soon the United States and China might limit their emissions, no matter what treaties the world's governments might one day agree upon. And as temperatures continued rising, this additional global warming would drive additional climate change: harsher hurricanes, deadlier wildfires, more epidemics.

By now I was passing Parliament, threading my way through crowds of tourists and office workers dashing out to lunch. King had said we were locked in to twenty-five more years of warming, but fifty years seemed more plausible. The reason was partly that, as I learned later in my reporting, other scientific analyses indicated that the climate system's inertia would keep temperatures rising after a global emissions halt for thirty to forty more years, not the twenty-five years King had cited. A second reason was that halting carbon dioxide emissions overnight is impossible: it would mean turning off most of the world's power stations, factories, vehicles, and other essential infrastructure—a recipe for chaos and suffering. Like it or not, fossil fuels were essential to our current social organization; it would take time to shift to alternatives. Historically, such shifts—from wood to coal in the nineteenth century, from coal to oil in the twentieth—had taken about fifty years. Even if we managed the task in half the time, we still faced at least fifty more years of intensifying summer heat, dwindling water supplies, and persistent droughts like the one then fueling civil war in Darfur. Lester

Brown's warning back in 1990 — that if we didn't reverse global warming within the next ten years, it could become irreversible — began to look disturbingly prescient.

Soon I had crossed the Westminster Bridge and begun heading down the far bank of the Thames. A large Ferris wheel, known as the London Eye, stood just ahead. I heard children laughing and shouting as they waited for the Eye's mechanical arms to lift and wheel them high above the bustling city. I was still a new father at that point, and it took these children's cries to remind me that I had a child of my own now. The words burst from my mouth before I knew it: "Chiara has to live through this."

It was a staggering realization. My infant daughter did not know how to walk or talk yet, but some fundamental facts about her future seemed already determined. Twenty years from now, when I hoped Chiara would be finishing college and preparing to make her way in the world, average global temperatures would still be rising, unleashing yet more powerful impacts. And temperatures and impacts were bound to keep increasing until at least 2050, when Chiara would be almost as old as I was now.

True, higher temperatures will have positive as well as negative effects. For example, as climate contrarians such as Danish statistician Bjørn Lomborg like to emphasize, fewer people figure to die from winter cold. But such positive effects will be dwarfed by negative ones, according to the vast majority of scientific analyses, including the IPCC's reports. It is also true that there is considerable uncertainty about the scope and timing of climate change impacts. Scientists find it especially difficult to determine the probability of the most extreme scenarios, such as the total melting of the massive Greenland and West Antarctic ice sheets or the shutdown of the Atlantic Ocean's thermohaline circulation — popularly known as the Gulf Stream — whose warm currents give Europe its temperate climate. But the practical consequences of such extreme events — an estimated forty feet of sea level rise if those ice sheets melt, a near-polar chill descending on Europe if the Gulf Stream shuts down — are so grave that they command concern. "The odds of some of the extreme scenarios may be only 10 to 20 percent, we're not sure," said Stephen Schneider, a professor of biology at Stanford University who was one of the first scientists to raise concerns about global warming in the 1970s. "But it's crazy to run those kinds of risks. The odds of your house catching fire are a lot less than 10 percent, but you wouldn't think of going without fire insurance."

The fundamental point is that my infant daughter would be growing up under global warming for the rest of her childhood and coping with climate change for the rest of her life. Under the circumstances, it wasn't just Chiara's physical safety I worried about; her emotional well-being was also at risk. As she got older, how would she cope with knowing that the climate around her would become less and less hospitable over time? How would that make her feel about her future, about perhaps having kids herself someday?

Staring down at the Thames, I felt stunned, heartbroken, but also deeply angry. Of course Chiara was not the only one at risk; every child on earth faced a version of the same fate. My fear mingled with a sense of personal failure, for my daughter and her generation were locked in to the very future that I and many other people had spent years trying to prevent. Now, it seemed, time had run out on all of us who had tried to halt global warming before it did serious damage.

But there was more to it than that. True, the premature arrival of climate change was partly a matter of bad luck. Even scientists as outspoken as Hansen were surprised by its speed. "The impacts we're seeing today weren't expected until late in this century," he later told me. Nevertheless, humans had played a decisive role.

Our collective failure to take action against global warming had been a conscious decision, a result of countless official debates where the case for reducing greenhouse gas emissions was exhaustively considered and deliberately rejected. Voices of caution had repeatedly been overpowered within the halls of government, in the media, and in the business world. Bankrolled by the carbon club lobby, to borrow author Jeremy Leggett's term—the energy and auto companies that profited from carbon emissions—opponents of taking action had confused the public, politicians, and the media with false or misleading information while also pressuring governments not to act.

Covering the climate story during the 1990s, I had often wondered about the deniers' motivations. Did they sincerely doubt the scientific case for man-made global warming? Or were their attacks rooted in an allegiance to continued burning of fossil fuels? Years later, an answer emerged after a lawsuit pried loose internal documents of the Global Climate Coalition. It turned out that the coalition's own scientific advisers had informed its leadership in 1995—two years before the carbon lobby led the fight against the Kyoto Protocol—that the science behind man-made global warming was "well established and cannot be denied." The coalition's board of directors responded by ordering their scientists' judgment removed from the coalition's public statements.

In short, the carbon lobby knew perfectly well that global warming posed real dangers, but it chose to deny those dangers and disparage anyone who sought to bring them to public attention. The lobby put its immediate economic interests ahead of humanity's future well-being. By devoting enormous financial resources and political muscle to blocking limits on greenhouse gas emissions, the carbon lobby in effect insisted that humanity bet its survival on the possibility that David King, James Hansen, and hundreds of other scientists were either lying or wrong about the dangers of climate change. Now, in October 2005, it was becoming clear that scientists had actually underestimated the danger. Humanity had lost the bet. Climate change had arrived a century sooner than expected, and future generations were no longer the only victims. My daughter and her peers around the world were now at grave risk as well.

As a father, I rebelled at what all this implied for my little Chiara's future. So there beneath the London Eye, I made a silent vow: to find a way, if one existed, for Chiara and her generation to survive the challenges ahead. Using my journalistic skills, I would investigate how bad things were likely to get, how soon. What would Chiara's community in northern California look like after ten, twenty, fifty more years of climate change? What were our civilization's chances and options for reversing global warming? Could we do so soon enough to avert what the IPCC had delicately called "the worst scenarios" of climate change, including an eventual sea level rise of eighty feet—enough to put most of civilization underwater? I also hoped to discover ways to cope with the heat waves, droughts, sea level rise, and other impacts that were now locked in over the coming decades. Could sufficient protections against these impacts be put into place? Above all, what steps were needed to turn these twin imperatives—to reverse but also to survive climate change—into practical realities? In short, what had to happen for my daughter and her generation to live through the storm of climate change?

Living Through the Storm

The first thing that struck me . . . was the magnitude of the risks and the potentially devastating effects on the lives of people across the world. We were gambling the planet.

—SIR NICHOLAS STERN, British economist, House of Lords

CHIARA AND I BEGAN reading fairy tales together long before she could understand the words or even focus her eyes on the pages. She was a week old, just released from her ordeal in intensive care, and normal things felt almost magical. It was bliss to sit in a rocking chair, cradle her tiny body against mine, and lull her to sleep with *The Three Billy-Goats Gruff, The Adventures of Peter Pan,* or *The Hobbit.* And so began our ritual. Chiara and I would read books together every night before bed and again the first thing the next morning, when we slipped downstairs early to give her mother some much-needed extra rest. We read fairy tales, nursery rhymes, picture books, Italian books, even adult nonfiction (the words didn't matter to Chiara at that point; it was enough for her to hear my voice). As the days became weeks and months, Chiara grew to adore books and the stories they contained. And her father came to understand that fairy tales offer valuable lessons to children and adults alike in the face of global warming.

Found in almost every culture, fairy tales are some of the oldest, best-loved stories on earth. They are passed down through generations not only because they amuse children (and help parents get them to sleep) but because they offer comfort and inspiration. In *The Uses of Enchantment,* psychoanalyst Bruno Bettelheim argued that fairy tales enable children to make sense of the world around them and to face the fact that "a struggle against severe difficulties in life is unavoidable, is an intrinsic part of human existence." But, Bettelheim continues, "if one does not shy away, but steadfastly meets unexpected and often unjust hardships, one masters all obstacles and at the end emerges victorious."

The first fairy tale Chiara fell in love with was *The Nutcracker*. She was about eighteen months old when she developed an obsession (and believe me, *obsession* is the word) with Tchaikovsky's magnificent score of E. T. A. Hoffmann's Christmas tale. Though she had only just begun to talk in full sentences, she insisted on hearing the story and music again and again. The plot is simple: At a Christmas party, Clara is given a nutcracker by her godfather, an inventor with a hint of magic about him. Clara falls asleep under the Christmas tree, clutching the toy. She awakens at midnight to see that the nutcracker, now grown as large as she, has come under attack from an army of giant mice, led by a king with seven heads. Just as the king is about to slay the nutcracker, Clara leaps into the fray and kills the mouse with a well-aimed hurl of her shoe. Her gesture transforms the nutcracker into a handsome prince, who shows his gratitude by inviting her to his kingdom, the Land of Sweets, where they live happily ever after.

After seeing *The Nutcracker* ballet onstage, Chiara began acting out the story at home. She invariably cast herself as Clara; her mother or I was assigned to play the godfather, the prince, or both. One day, after she and I had played the game for about the three hundredth time, I got distracted. To my half-listening ears, the music seemed to indicate the start of the battle scene, so as the prince I began to brandish my sword. A puzzled look appeared on Chiara's face. It took her a moment to realize that her father was confused. She looked up and carefully explained, "No, Daddy. It is still the party. The danger is not here yet."

The party, so long and pleasurable, that gave rise to global warming is indeed still under way. Despite years of warnings about overheating the atmosphere, we humans are still merrily riding in cars and airplanes, building pipelines and power plants, gobbling meat, clearing forests, expanding our houses and suburbs, and doing a thousand other things that emit the greenhouse gases that cause the problem. There has been a lot of talk about going green, but the economies of most nations are still based on burning oil, coal, and other carbon-based fuels, so emissions continue to increase. Meanwhile, the party gets more crowded and raucous by the day, as global population swells, the wealthy pursue ever more luxurious lifestyles, and the poor yearn for their own taste of the comforts fossil fuels can provide.

If most of us nevertheless seem in no hurry for the party to stop, the second half of Chiara's statement suggests why: the danger is not here yet, at least for most of us. The majority of the world's people have not been hit by climate change yet; it has not cost us a house, a livelihood, or a loved one. Sure, we may feel nervous about the recent erratic weather, we may feel disturbed by news reports of distant tragedies, but our daily lives continue pretty much as before. And so the party continues.

For millions of less fortunate people, however, indifference to climate change has become an unaffordable luxury. For them, the danger is now.

While visiting Bangladesh for this book, I met a little girl who was almost exactly Chiara's age. Her name was Sadia, and her father was the unofficial mayor of a village that was literally disappearing beneath his feet. The village, Antarpara, used to straddle the mighty Brahmaputra River. Like most of the rivers that course through Bangladesh, the Brahmaputra originates in the snowpack of the Himalayan mountains. But rising temperatures were now melting the snow faster and, along with stronger monsoon rains, boosting the river's volume. No one could say for sure that the excessive flooding was caused by global warming—after all, Bangladesh has a long history of flooding. But the flooding of Antarpara was certainly consistent with what scientists projected as global warming unfolded: faster glacial melting and more volatile monsoon rains.

"You cannot definitively attribute any single extreme event to climate change, but the overall pattern is clear," said Saleemul Huq, a Bangladeshi biologist who directed the climate change program at the Institute for International Economics and Development in London and who had invited me to his native country. "In Bangladesh, we know very well what a 1-in-20-years-size flood looks like. We've had them for centuries. But in the last twenty years, we've had four floods of that magnitude: in 1987, 1988, 1995, and 2005. This suggests we have entered a new pattern where we get a 1-in-20-years event about every 10 years. This is something we have to worry about now, not in the future."

Anisur Rahman, the mayor of Antarpara, was a broad-shouldered man who wore a dirty blue shirt and tattered rubber sandals. As we stood by the bank of the Brahmaputra, gazing out at the sluggish, silver-white current, he told me, "This river comes from India. For some reason, the water in India is increasing, so the floods here are bigger. They are sweeping away our houses, even the land beneath them. There were 239 families in this village before. Now we are 38 families."

Clustered around the mayor as we talked were dozens of villagers, mainly women in cheap bright saris—lime green, sky blue, scarlet—with skinny children clinging to their necks. "I have had to move my house seven times in the last twenty-eight years," said Charna, a haggard mother of two. "I used to live over there," she said, gesturing toward the middle of the river, "but floods washed the land away and I had to move here."

Later, when I bade the mayor goodbye, he was holding his daughter in his arms. Sadia was a pretty, solemn little girl, about eighteen months old. She was the mayor's first child, and he definitely wanted her to go to school one day, but it would not be in Antarpara. "By the time she is old enough," he explained, "this village won't be here."

There is a terrible injustice at the heart of the climate problem: climate change punishes the world's poor first and worst, even though they did almost nothing to bring it on. After all, they cannot afford to drive gasoline vehicles, fly in airplanes, eat much meat, or inhabit the climate-controlled buildings that are the principal contributors to global warming. "Eighty percent of global greenhouse gas emissions come from the richest 20 percent of the world's people," said Saleemul Huq. "The poorest 20 percent of the world's people are responsible for less than 1 percent of emissions. But because of their lack of resources, they will probably account for 90 percent of the deaths those emissions cause. This means that climate change is no longer just an environmental or energy or economic problem. It is also a justice problem."

"You'll Remember How Nice Summers Used to Be"

Even for the rich, climate change is now a matter of self-interest. "I attended a conference recently and found myself talking with an executive of DuPont, the chemical company," said Chris West, the director of the UK Climate Impacts Programme, a British government agency that educates local governments, businesses, and individuals on how to manage the impacts of climate change. "[This executive] told me about all the green initiatives that DuPont had launched—shrinking its carbon footprint, reducing its toxic emissions, just treating the environment better in general. 'Jolly good,' I said. 'But is DuPont also prepared for how the environment might treat you?' He didn't know what I was talking about. I asked how many facilities his company had around the world. 'About three hundred,' he said. I asked how many of them were located in floodplains. He didn't know. I said, 'Don't you think you should?'"

As we begin the second decade of the twenty-first century, every person on earth finds himself or herself in the same boat as that DuPont executive. Like the executive, we are largely unaware of what is about to hit us, even as we congratulate ourselves on our blossoming environmental awareness. Many of us have heard about global warming and want action taken against it. But few of us have reckoned with the inconvenient truth that climate change is going to keep coming at us no matter what for a long time. We do not realize that serious climate impacts are inevitable in the years immediately ahead. We have not considered how harsher heat waves, melting snowpacks, and other inevitable climate impacts will affect our work, homes, children, and communities; much less have we taken steps to reduce our vulnerability.

Don't you think we should?

"The point we have to get across to people is that the future is not going to be like the past. It's human nature to assume it will be, but with climate change that's no longer true," said Kris Ebi, an independent scientist who began analyzing global warming while working for the U.S. electric utility industry and later coauthored a chapter of the *Fourth Assessment Report* about health impacts. "I do a lot of speaking at colleges and universities, and even there this message hasn't gotten through," added Ebi, who has two adult daughters. "I told one class, 'When you're my age, you'll think back to how nice summers used to be. Summers in the future will be a lot less comfortable than today.'"

How did the students respond? I asked.

"They didn't say much, but their eyes got very big," Ebi replied.

Fear of climate change is only natural, and it is perhaps inevitable that some people take refuge in denial. One father I met in San Francisco, a city proud of its green consciousness, told me that he deliberately avoided news about climate change—it was too depressing, especially when he thought about the implications for his kids, aged seven and four. "I think people my age will be all right," he said. "Things will be tolerable for the next twenty years or so. But our kids are screwed."

Avoiding unwelcome truths may be standard procedure for human beings, but it isn't much of a survival strategy. If there is even a slight possibility of improving our children's chances of coping with what lies in store, how can we choose denial? We wouldn't do that if our child were diagnosed with a life-threatening illness; we would face the awful facts, find the best doctors we could, and pursue every possible treatment option. When Lisa Bennett, a Bay Area mother of two young boys, awoke to the dangers of climate change, she felt compelled to take action. She later explained, "I began to think it a bit crazy that I attended to every bump and scrape on my children's little bodies and budding egos but largely ignored the threat likely to put sizable areas of the world, including parts of the coastal city where we live, underwater within their lifetime."

To borrow again from fairy tales, it is facing the dragon, as scary as that may be, that calls forth the heroes who deliver victories. "The baby has known the dragon intimately ever since he had an imagination," observed the writer G. K. Chesterton. "What the fairy tale provides for him is a St. George to kill the dragon." Often the heroes who kill dragons are ordinary people, as frightened as anyone but impelled to do the right thing. In *The Nutcracker*, Clara must attack the seven-headed mouse king in order to save her beloved nutcracker. In *The Wizard of Oz*, Dorothy and her companions must bring back the broomstick of the Wicked Witch of the West before their wishes are granted. In the Harry Potter series, the young hero must confront and defeat his parents' murderer.

Now, in the struggle against climate change, we need thousands of ordinary heroes to step forward and fight for our future.

Happily, there are genuine reasons for hope. Not only do we know what it will take to stop global warming, but most of the necessary technologies and practices are already in hand. Best of all, putting these tools to work could actually strengthen our economy, improve our quality of life, and make money, lots of it.

Ironically, one of the biggest profitmakers is a company that later caused the largest environmental disaster in U.S. history, the BP oil gusher that fouled the Gulf of Mexico in 2010. But in 1999, under different leadership, BP had invested in energy efficiency, which is by far the quickest, most lucrative way to reduce greenhouse gas emissions. BP invested $20 million to install more efficient light bulbs, motors, and operating schedules in the company's refineries, offices, and workplaces. Over the next three years, those efficiency improvements lowered BP's energy bills by $650 million. Thus the company's original $20 million investment yielded a profit of $630 million—a stunning thirty-two-fold return on investment. Even organized crime doesn't enjoy those kinds of profit margins.

Plenty of other corporations are following the same path, and so are forward-thinking governments. In Germany, Chancellor Angela Merkel's conservative government has subsidized energy efficiency investments that were initially devised by the left-of-center Green Party. Every year, the German government funds the renovation of 5 percent of the nation's pre-1978 housing stock, covering the up-front costs of installing more efficient insulation, heating, and electrical systems. The program is widely regarded as a win-win-win. The annual 1.5 billion Euros in subsidies are recouped through lower energy costs. Greenhouse gas emissions are reduced. And perhaps most important for a nation struggling with high unemployment rates, the program generates thousands of jobs for construction workers, jobs that by their nature cannot be sent abroad.

In the United States, the state of California boasts comparable achievements. Under the leadership of Governor Jerry Brown in the 1970s, California launched a sustained effort to improve energy efficiency, especially regarding electricity use. We'll discuss specifics in a later chapter, but the results have been remarkable. California's electricity consumption today is roughly the same as thirty years ago, even as the state's population and economy have grown tremendously.

California, Germany, and BP are but three examples of the larger truth: if we're smart, the fight against climate change can repair, not ruin, our economies. Renovating our homes, workplaces, farms, transportation, and other systems to run on low-carbon energy sources will cost money up front, but it will create jobs, spur innovation, and boost profits over the long term. Installing the protections needed against heat waves, sea level rise, and other future climate impacts could likewise stimulate enormous amounts of economic activity, especially for the construction industry and other labor-intensive sectors. Indeed, the green economy is shaping up as the largest growth field of the twenty-first century; a 2009 study by the HSBC Bank calculated that the global green economy will grow from a $500 billion market today to a $2 trillion market by 2020. Germany and China, the world's two leading export powers, clearly recognize this opportunity and are moving quickly to seize it; the jury is still out on the United States.

Energy efficiency is not a silver bullet, nor can it forever neutralize the effects of billions of people consuming more and more all the time. If the consumerism of the rich, the population increase of humanity as a whole, and the ceaseless growth imperative of modern capitalism continue unchecked, their impacts will cancel out the gains of even the most ambitious efficiency programs. Nevertheless, improving efficiency is a crucial first step. Because it is so profitable, it can generate funds to develop and deploy the solar panels, carbon-neutral buildings, protective seawalls, and countless other technologies that are needed both to reduce emissions and to cope with the unavoidable impacts of those emissions. And because it is fast-acting, energy efficiency can buy us time to deploy these technologies while we wrestle with the deeper challenges of taming consumerism, limiting population growth, and reorienting our economies from material growth toward alternative measures of well-being.

Another piece of good news: climate change does not necessarily doom the poor. The most hopeful story I uncovered while researching this book was in Africa, the continent scientists say will be hit hardest by climate change. In the sun-baked Sahel, I talked with illiterate farmers who did not know the term *climate change* but were adapting to it nonetheless. To capture rainfall and rejuvenate soil fertility, the farmers were growing trees amid their fields of millet and sorghum. With little outside funding, their techniques had spread from village to village across vast areas of Niger, Burkina Faso, and Mali, with remarkable results: despite enduring some of the hottest, driest weather on earth, greenery has returned to more than 12.5 million acres of land. Underground water tables have risen. Crop yields have doubled and tripled. To be sure, life is still hard in the Sahel, and it is bound to get harder still as temperatures rise further in the years ahead. But the region's farmers are by no means surrendering in the face of climate change, and they may yet survive it if the outside world does its part and slashes greenhouse gas emissions.

Global warming is not the only reason our civilization must shift to low-carbon energy sources: there is also the threat of "peak oil." As recently as five years ago, the theory of peak oil—which holds that humanity has already consumed half of the oil on the planet—was derided as nonsense from the fringe. No longer. As stalwart a member of the energy establishment as James Schlesinger, a former director of the CIA and secretary of the U.S. Departments of Energy and Defense, said in 2007, "The debate is over—the peak-ists have won." There is still lots of oil to be had on this planet, but it "will get harder and costlier to find," Ronald Oxburgh, the former chairman of the British arm of Royal Dutch Shell oil, told me. (Peak oil is one reason BP was drilling so deeply in the Gulf of Mexico in the first place.) Meanwhile, global demand for petroleum continues to climb. If and when global demand outstrips supply, analysts warn, the imbalance could bring debilitating shortages, soaring prices, crashing economies, resource wars, and social breakdown. The car-dependent lifestyle that millions of Americans (and growing numbers elsewhere) take for granted will become impossible. Fatih Birol, the chief economist of the International Energy Agency, is another insider worried by the approach of peak oil. "We should leave oil before it leaves us," Birol wrote in 2008.

Make no mistake: going green at the speed and scale needed to defuse global warming and escape peak oil will not be easy. We will have to abandon old ways of thinking, confront powerful interests, spend large amounts of money, adjust our material appetites, and stay focused on the mission for many years to come. But there are unsung heroes all over the world who are already working to make these changes a reality; you will meet some of them in this book. They deserve our help.

The Double Imperative of the Climate Fight

Chiara happened to be born at a momentous turning point in human history. What I call the first era of global warming began on June 23, 1988—the day NASA's James Hansen told the U.S. Senate that man-made global warming had begun. Although a handful of insiders were worried before then, it was Hansen's testimony—and the attention it received after the editors of the *New York Times* ran the story on page 1—that put the world on notice that civilization's future is at risk. Global warming quickly became a common phrase in news bureaus, government ministries, and living rooms around the world. When a top scientist at the agency that put a man on the moon warns that trouble is brewing, attention must be paid.

As emissions kept growing, climate change went from being a distant theoretical danger to a punishing current reality. This shift took place sometime around the turn of the twenty-first century (scientists are still determining the exact date), inaugurating the second era of global warming. The battle to prevent dangerous climate change was now over; the race to survive it had begun. If humanity is to win this race, the essential first step is to change the way we think about climate change. The climate problem has undergone a paradigm shift; we humans must now make a paradigm shift of our own.

Today, in the second era of global warming, humanity faces a double imperative. On the one hand, we must reverse global warming, and quickly—before the climate system passes tipping points that could trigger irreversible warming. "We're about ten years from a point of no return," Al Gore told me in 2006. "But we still have time to slow the rate of warming and thereby buy time for the introduction of revolutionary technologies and practices that could reduce emissions enough to avoid the worst impacts of climate change." Yet even as we strive to lower the global thermostat, we must also go beyond this traditional definition of climate action. We must take steps as well to prepare our societies for the serious climate impacts that are already in the pipeline. In short, we have to live through global warming even as we halt and reverse it.

At present, this double imperative remains unrecognized by many of us, whether we are individuals, communities, businesses, or governments. Over the past few years there has been an explosion of concern about global warming. But if awareness is high, understanding remains low, in rich and poor countries alike, among both the general public and policymakers. To hear most politicians, corporate advertisements, media reports, and even environmental groups tell it, fighting climate change is all about shifting to cleaner energy sources (and—a distant second—changing farming and forestry practices). If we switch to solar, wind, and other low-carbon energy sources, we can "Stop Global Warming," to quote one oft-heard slogan, in the same way we turn off a car engine. But few people seem to recognize how quickly this shift must be made, or grasp how substantial the impacts will be in any case. A better analogy is to imagine that our civilization is traveling in a train, heading downhill, picking up speed, and approaching a landscape obscured by storm clouds. We can hit the brakes by reducing greenhouse gas emissions, and we must. But the train's momentum ensures that it will be a long time before we actually come to a halt, and before we do, we will cross a great deal of unknown territory.

In triggering climate change, humanity has unwittingly launched an unprecedented planetary experiment. Because this experiment has never been run before, and because it involves extremely complicated systems, knowing exactly how it will turn out is impossible. What we do know is, we are pushing the earth's climate system well beyond its normal limits. The past 250 years of industrialization have increased the amount of carbon dioxide in the atmosphere to 390 parts per million—the highest level in the last 800,000 years, and probably in the last 20 million years. We know further that this increase has not only caused global warming but contributed to concrete examples of climate change, such as the 2003 heat wave in Europe, and that such impacts will intensify in the future. Nevertheless, there are many specifics we do not know. For example, the years ahead are expected to bring an increase in the frequency of extra-strong hurricanes. But exactly when and where they will strike, no one knows.

This lack of scientific certainty is no cause for reassurance. From the beginning of the climate debate in the early 1990s, those opposed to taking action have used the lack of certainty to argue against reducing greenhouse gas emissions. Why damage the economy, they asked, unless we are sure such reductions are required? The developments of the last ten years reveal the recklessness of that argument. Opponents ignored that scientific uncertainty can cut both ways—yes, things can turn out better than expected, but they can also turn out worse.

That simple piece of common sense is the basis of the precautionary principle. A cornerstone of modern environmentalism, the precautionary principle holds that policymakers should err on the side of caution when making a decision that carries apparent but uncertain risks. Put differently, the absence of definitive proof that a given activity is dangerous does not prove it is safe. But the precautionary principle has been ignored in the battles over climate policy. Alas, real-world experience and additional scientific observation and analysis have now demonstrated the folly of this course. The climate system has turned out to be much more sensitive to global warming, much more prone to human disruption, than anticipated.

"In the last few years, we've gotten strong hints that we've underestimated this problem, not overestimated it," said Peter Gleick, founder of the Pacific Institute in Oakland, California, and one of the world's leading experts on water policy and climate change. "Scientists can be conservative when it comes to drawing controversial conclusions, especially when they know they will be attacked for them. In the water area, we're seeing many developments consistent with the worst scenarios projected for future climate change. For example, we're in the middle of a very extreme drought in the southwest and the southeast of the United States. We're not certain yet that climate change is causing these extremes—history shows that the hydrological cycle is characterized by extremes—but it is entirely possible."

Sir Nicholas Stern famously remarked in his 2006 study of the economics of climate change that climate change represented "the greatest and widest-ranging market failure ever seen." Prices, government regulations, and other market forces had not only failed to prevent climate change, Stern pointed out, they had encouraged greenhouse gas emissions to grow and grow. Now, we can say that climate change also represents the greatest and widest-ranging failure of the precautionary principle ever seen. In the face of uncertain but potentially catastrophic consequences from increasing emissions, our economic and political leaders chose to pursue business as usual, presuming that the risks would turn out to be manageable. The coming years will instruct us about how manageable they actually are.

It is often supposed that rich societies and individuals will find it relatively easy to adapt to climate change; their money and technological prowess, goes the argument, will enable them to counter harsher heat waves with more air conditioning and stronger storms with sturdier seawalls. Leave aside for the moment the fact that this assumption ignores the plight of the world's poor, who amount to roughly half the people on earth. My research for this book has convinced me that even wealthy, technologically advanced societies will find it enormously challenging to defend themselves. The climate impacts that are already in place are so large, pervasive, and interlocking that they will tax our adaptive capacity to the maximum, especially because we will be confronting them at the very time we are grappling with peak oil and global economic disorder, not to mention the necessity of reversing global warming before its impacts increase from the "merely" grave to the outright apocalyptic.

Over the next fifty years, climate change will transform our world in ways we have only begun to imagine. Humans have changed the weather on this planet, and that will change everything: from how we grow food and obtain water to how we construct buildings and fight disease; from how we organize economies and control borders to how we manage transportation systems and deploy armies; from how we write insurance and produce wine to how we talk with our children and plan for the future.

By no means is climate change the only threat to our civilization's future, but it tends to intensify other outstanding threats, whether military, economic, or environmental. Military experts call climate change "a threat multiplier," to quote a 2008 report by the European Union's two top foreign policy officials. Climate change will worsen existing conflicts over water supplies, energy sources, and weather-induced migration, the report warned, potentially "overburden[ing] states and regions which are already fragile . . ." Economic prosperity is also endangered.

Approximately 25 percent of the gross national product of the United States is at risk from extreme weather events, according to the American Meteorological Society and the American Geophysical Union.

Global warming and climate change also undermine the ecosystems that make human life possible on this planet, ecosystems that our civilization has already placed under extreme stress. In particular, global warming and climate change hasten the loss of plant and animal species, which is arguably the single most worrisome global environmental trend after climate change itself. Already, temperatures and climates are shifting too fast for many species to adapt, especially in the face of rapid habitat loss, which has been the primary cause of species loss to date. Writing in *Nature* in 2006, nineteen of the world's leading biodiversity scientists warned that climate change alone could lead to the extinction of between 15 and 37 percent of all species by 2100. Need one add that such a massive loss of other species raises the odds that humans will also go extinct sooner rather than later?

Indeed, over the course of writing this book, I have come to see the climate crisis as a major evolutionary test for our civilization and perhaps our species. Like all such tests in the long, long history of evolution, it will be the individuals who can adapt to the new conditions best who will survive and prosper. Those who cannot adapt, meanwhile, will perish, perhaps not immediately but before very long.

The Third Era of Global Warming

The inevitability of fifty more years of rising temperatures and their associated impacts is the great unfolding story of our time. The implications haven't sunk in yet to most people, but it won't be long; reality is a powerful teacher.

Yet reality is also a work in progress. Temperature rise and the physical effects it causes may be inevitable, but how humans react is up to us. There is still time, if we hurry, to enlarge our vision of how to cope with climate change—to recognize that we must not only reverse it but also adapt to it. Only such a shift in thinking and action can give our children, future generations, and the natural world we all depend upon a fair chance of living through the gathering storm of climate change.

With Chiara, Sadia, and the rest of their generation foremost in mind,

I aim in this book to call attention to the new realities of climate change; to provide a hopeful but realistic picture of the changes that lie in store over the next fifty years and beyond; and to identify the best steps to both reverse global warming and adapt to its impacts. Some of these steps you can take as an individual. Others can be taken only by governments. Still others fall to the private and civil sectors. Individuals can plant trees, conserve water, and do a thousand other valuable things, but it is governments that must build seawalls and set overall energy and economic policies. It is corporations that must quit dirty fuel sources like coal and embrace alternatives like efficiency and solar.

Chiara, Sadia, and their peers belong to what I call the climate change generation: the nearly 2 billion people who have been born since the first era of global warming began in 1988. In the years ahead, the young people of this generation must not only learn to live with the climate disruptions their elders have set in motion; they must also bring about the green revolution that is our best hope against descending into outright climate chaos. "I was giving a talk recently to a class of high school kids and I told them that in the next forty years, because of global warming and other environmental problems, everything about our society is going to have to change," said Richard Louv, author of *Last Child in the Woods*. "I told them we need a new energy system, [that system] is already beginning, and they need to build it out. I told them we need a new agricultural system, [that system too] is already beginning, and they need to build it out. The kids were rapt, which surprised me, because I'm not much of a public speaker. Afterwards the teacher told me it was because I'd told them something hopeful about the environment, and they never hear that. Fear may motivate some kids to get involved, but most need to hear a message of hope."

Besides hope, the youth of the climate change generation need the help of their parents and grandparents—the grownups who run today's society and have the greatest immediate power to change its course. Previous generations, Hansen notes, "did not realize the long-term effects of fossil fuel use. We no longer have that excuse." Taking action on climate change, I would argue, has now become part of a parent's job description, no less vital than tending to your child's diet, health, or education. Just as no responsible parent would encourage a child to smoke cigarettes, so parents henceforth should be reducing their families' carbon footprint (and pressing governments and corporations to do the same) while also strengthening their households' and communities' resilience to climate impacts.

If all goes well, the next fifty years may be remembered as the second of three eras of global warming—a bridge between the first era of discovery and delay and a third era of deliverance and survival. If governments, communities, institutions, and individuals can rise to the challenges of the next few years, the second era of global warming could be a time of victory and redemption. By 2050, humanity might have slashed greenhouse gas emissions, limited temperature rise, and put in place protections against many climate impacts. There will, alas, still be losses; our collective failure to act sooner means we cannot save every person and place we would like. But we can keep the losses to a minimum if we act boldly. In that case, humanity might enter a third era of global warming. The inertia of the climate system ensures that temperatures will remain high in this third era; sea levels will still be rising, other impacts still unfolding. But the worst might be past. And buoyed by the lessons learned, humans might begin a new kind of existence on this planet, one based on equity and sustainability. Many of today's adults will not live to see this third era, but our children and grandchildren could, and that is reward enough.

Like a fairy tale, this book contains heroes and villains, dangers and triumphs, tests and judgments. The story remains unfinished, however. The ending depends, as in most fairy tales, on the choices made by the characters, which is to say by each one of us. George Woodwell, a biologist who cofounded the Woods Hole Research Center in Massachusetts, articulated the basic choice at the Chicago Humanities Festival in 2007. "If today's trends continue much longer, this earth will become a hell," Woodwell, himself a grandfather, told the crowd. "But we don't have to build hell. We can tell our grandchildren instead how they can make the new world we need. At my institute, we live in a building that does not use a flame. It has gotten its electricity from solar panels for the past twenty-four years. We can do this, if we want to. It all depends on what future we decide to build."

2 | Three Feet of Water

Nature, to be commanded, must be obeyed.

—FRANCIS BACON

AN HOUR'S DRIVE north of San Francisco is the tiny coastal town where our family spent the first few years of Chiara's life. As I stand at the kitchen sink filling the kettle for morning tea, the view out the window is of nothing but trees and sky all the way down to the Pacific. We can't see the ocean—it's half a mile away and pines and cedars block the sightlines—but we do hear it. When the wind is right and the waves are strong, the crashing of sea against shore is unmistakable. Sometimes the sound is faint yet distinct, like a memory you didn't know you had; other times it's loud and boisterous, like a subway train roaring past. Never is it boring.

It takes five minutes to walk from our front door to the beach, fifteen minutes if Chiara comes along. The last thirty yards lead down a trail that offers a view across the sea of one of the last functioning lighthouses on the Pacific coast. The beach is mainly rocky at our end, covered by eroded shale that has trickled down from cliffs that tower fifty to a hundred feet into the air. Low tide exposes a black, slippery reef where Chiara loves to explore tide pools that teem with mussels, baby crabs, and other aquatic life. At least twice a week she goes to play at this beach or the sandy one in town, where she quickly wiggles out of her shoes and runs off, blond curls blowing in the wind, to splash in the chilly surf and build sandcastles.

Our town is perched on the end of a peninsula; a lagoon fed by the ocean separates us from the mainland. Three days a week, we drive Chiara to the other side of the lagoon for preschool. The trip takes twelve minutes: first along a two-lane country road that is the only way in and out of town, then on a two-lane coastal road that hugs the opposite side of the lagoon. The lagoon is part of a nature reserve, one of the last remaining spots where migratory birds can stop to feed and rest during their travels up and down the Pacific coast, so we often see lots of wildlife. Depending on the season, there may be egrets, ducks, terns, sandpipers, or brown pelicans swooping through the sky, foraging in the muddy shallows, or simply resting. The kings of leisure are the sea lions. All but motionless, they lie in long rows on barely submerged sandbars like giant cigars drying in the sun.

Chiara's school is located at the far end of the lagoon, on a grassy patch of land at the base of the coastal ridge. From the playground you can look back across the water and, if it's not too foggy, see our town about a mile away, including the wharf where sailboats from San Francisco used to land in the 1850s, when the town was first established during the Gold Rush.

Chiara sees the ocean in one form or another nearly every day here, and she loves it without reservation. Once I asked what part of the ocean she liked best: The water? The beach? The tide pools? Cocking her head, she looked at me as if those were pretty silly questions. "All of it, Daddy," she replied. "I like all of it." Not only does she like it, she feels at ease with it. After all, it's what she has grown up with.

But for me, a farm boy who grew up back east, it's a new experience to live so close to the ocean. I am awed by its power and vastness, but what impresses me most is its relentlessness. Night or day, rain or shine, the ocean keeps coming. It never stops coming.

Now, because of global warming, it will not stop rising either. Even if greenhouse gas emissions were to fall soon and rapidly, sea levels would keep climbing for the rest of Chiara's lifetime and far beyond—indeed, for thousands of years. Higher temperatures cause seas to rise in two ways. First, they melt glaciers and polar ice into water that eventually flows into the ocean. Second, they warm the ocean, and since warmer water expands, the ocean's volume increases. There are long lag effects to both of these phenomena. The IPCC projects that even if global emissions had been capped in the year 2000, the temperature rise already locked in to the system would cause glaciers to shrink and polar ice to keep melting for hundreds of years and oceans to keep expanding for thousands.

"In climate change, as in comedy," says British science writer Mark Lynas, "timing is everything." Three feet of sea level rise over the next thousand years would be little cause for alarm. Three feet of sea level rise over the next *hundred* years—which is near the low end of what scientists now expect—will pose enormous challenges.

Unless seawalls and other barriers are installed, a sea level rise of three feet would bring catastrophic flooding to many of the world's leading cities, coastlines, deltas, and islands. This is especially so because along with sea level rise, climate change will also be causing stronger storms. Three feet of sea level rise will gravely affect an estimated 145 million people around the world, most of them in Asia. The world's chief financial capitals—New York, London, and Tokyo—are all highly vulnerable, thanks to their low-lying waterfront locations. I visited each of those cities for this book, as well as Shanghai, the epicenter of Chinese capitalism, where three feet of sea level rise would put a third of the city underwater. Mega-cities located in poor countries would be equally pressed and much less able to adapt; Manila, Jakarta, and Dhaka are the three considered most at risk in Asia.

In the United States, a mere two feet of sea level rise would put 2,200 miles of roads in Washington, DC, Virginia, Maryland, and North Carolina at risk of regular inundation, according to a 2009 report by the U.S. Geological Survey, the National Oceanic and Atmospheric Administration, and the Department of Transportation. Worldwide, approximately $3 trillion of assets are located at or below three feet above sea level, according to the *Stern Review*, an analysis of the economic implications of climate change published by the British government in 2006. The assets at risk include infrastructure crucial to modern society: water treatment facilities, power stations, railroads, highways, buildings, airports. Still more low-lying land is occupied by housing and agricultural activities, not to mention beaches, wetlands, and other vital ecosystems.

In theory it is possible to move or protect these assets, but doing so will be neither quick nor cheap. To build sea defenses around airports or to relocate coastal communities will require billions of dollars and decades of time. No one knows exactly how much time we have. Forget a hundred versus a thousand years: some scientists believe our civilization could experience three feet of sea level rise within the next *fifty* years. This is an extreme but by no means impossible scenario. Indeed, it is what the legendary insurance company Lloyd's of London has been told to expect by one of its scientific advisers, Professor David Smith of Oxford University, who projects that sea levels will rise 2 meters (6.5 feet) by 2100.

"We Have to Accept That There Will Be Losses"

The inevitability of considerable sea level rise is a defining characteristic of the second era of global warming. As such, it requires a corresponding shift in policy and behavior from individuals, governments, businesses, and civic institutions. From now on, we must not only pursue the traditional goal of climate policy—reducing greenhouse gas emissions—but also add a new focus on what climate scientists call adaptation.

When climate scientists use the word *adaptation,* they mean actions intended to reduce one's vulnerability to the impacts of climate change. For example, when a community constructs sea defenses to protect against hurricanes, or when a homeowner plants trees around the house to shade it from extra heat, that is adaptation. By contrast, scientists use the word *mitigation* to refer to actions that reduce the greenhouse gas emissions that cause global warming in the first place. For example, an electric utility that closes a coal-fired power plant and builds wind turbines is practicing mitigation. So is a commuter who takes a bus or bicycle to work rather than driving a car. In a sense, mitigation addresses the front end of the climate problem; by cutting emissions, it aims to limit the eventual increase in temperatures and the impacts they cause. Adaptation, on the other hand, addresses the back end of the problem:

it increases one's resilience to the impacts that previous emissions have already set in motion.

Adaptation will be especially important in Chiara's home state of California. Sea levels along the California coast are expected to rise between 3 and 4.5 feet by 2100, according to a landmark study that the state commissioned from the Pacific Institute in Oakland. Released in March 2009, the study warned that such sea level rise would endanger $100 billion worth of property and 480,000 residents, many of them economically disadvantaged people who would find it difficult to move or protect themselves.

One of the officials who pushed hardest for this study to be done was Will Travis, the executive director of the San Francisco Bay Conservation and Development Commission, which regulates activities in the bay. A few days before the study was released, I watched Travis deliver a presentation that was a rarity in the field: a discussion of climate change that was almost as amusing as it was alarming.

The occasion was a hearing of the San Francisco Public Utilities Commission, which manages water and power for San Francisco; Travis's audience included the PUC's five commissioners and about fifty members of the public. (Disclosure: Chiara's mother, Francesca Vietor, was one of the five commissioners.) A stocky, grandfatherly looking fellow, Travis began by noting that three feet of sea level rise by 2100 was the minimum to expect. To make the consequences vivid, he showed satellite images of San Francisco and nearby localities today, followed by images of how they would look after seas rose three feet. One casualty: San Francisco International Airport; the "after" image showed its runways, terminals, and access roads underwater. The audience gave an audible murmur.

"So," Travis continued in a chipper voice, "the solution to climate change is clear: fly Oakland." People chuckled. But then, showing Oakland's airport, he said, "Oops, maybe not." Oakland's airport too was underwater after three feet of sea level rise. People laughed again, this time a bit nervously. Travis went on to say that parts of Silicon Valley, which borders San Francisco Bay, would also be inundated, especially if one factored in storm surges: scientists had projected that by 2100 today's 1-in-100-year storms would be occurring once every ten years.

Travis's commission was collaborating with the Dutch government on a response plan, he said: "We'll have to build a lot of levees, levees strong enough to withstand earthquakes." Then he added the kicker: "But we shouldn't build levees everywhere. In some places, it may be best to remove existing developments"—here he showed a photo of tract housing—"and replace them with tidal wetlands, which are close to magic when it comes to coping with climate change." No one was laughing now.

As Chiara's father, I found it encouraging that a public official as influential as Travis was so engaged with sea level rise. Most impressive was to see Travis broach the great unmentionable in adaptation discussions: the fact that we can't save everything. His call to remove existing developments was an admission that even a community as financially and technologically blessed as the Bay Area, the epicenter of the Internet revolution and home to some of the richest people on earth, simply cannot protect every single place that is imperiled by climate change.

"We have to accept that there will be losses," Suzanne Moser, a scientist formerly with the National Center for Atmospheric Research and one of the first American experts to investigate the role of adaptation in climate policy, told me. "We can't do everything, everywhere. Even if we could financially afford to build a seawall around the entire continent, it wouldn't be the right way to go. It would be Fortress America, and besides, the ocean is just way more powerful. You can put as much money against the ocean as you want; eventually the ocean will win."

If losses are inevitable, said Moser, the human response should be twofold: minimize those losses by accelerating our mitigation efforts, and prioritize the losses—focus our adaptation efforts on saving the places and people that matter most. That sounds sensible in the abstract, but the reality promises to be messy. Who makes such decisions? Who pays? And who delivers the bad news to those deemed impractical to save?

Our beach town lies fifteen miles as the crow flies from the Golden Gate Bridge at the mouth of San Francisco Bay. (As it happens, tethered near the bridge is the oldest functioning sea level gauge in the United States—one reason the scientific data at Travis's disposal is so robust.) A few days after Travis's presentation, I took a look around our town to ponder how three feet of sea level rise would affect us.

Depending on how quickly seas rise, most of our town's houses would probably be fine, for they stand well above sea level. For example, the vertical drop between our front door and the beach where Chiara plays in the tide pools is about two hundred feet. The trouble for our town will be the roads. On the mainland side of the lagoon, the coastal road hugs the shoreline, and while the pavement is above the waterline, it's not above it by much—perhaps five feet in most places. In two spots, the road is just a couple of feet above the waterline, making flooding inevitable unless protection is installed.

There is a second way in and out of town: you can take the coastal road from the north, where the land is high. This would add a half-hour of travel time, but the real problem is that you still must take the country road the last two miles into town. That road sits well above the lagoon for most of its length, but its final half-mile passes through Gospel Flats.

Home to one of the first modern organic farms in California, Gospel Flats is a lovely piece of land, especially at harvest time, when hundreds of pumpkins sprawl by the roadside. We take Chiara there every Halloween with her cousin Lana to pick out jack-o'-lanterns and make apple cider from an old hand press. But Gospel Flats is barely a foot above sea level; the marsh begins just beyond the pumpkin patch. I've seen heavy rains turn the road there into a pond that only high-slung pickup trucks can navigate. So even one foot of sea level rise—which could plausibly occur in as little as fifteen years—would cause real difficulties. Three feet of sea level rise would make Gospel Flats pretty much impassable.

So what to do? In theory our town could build a causeway across Gospel Flats. But we have only 1,200 inhabitants—how would we pay for such a project? We could seek funding from the state government, but the state will be fielding plenty of similar requests from throughout California in future years, most of them from places bigger and more economically valuable than we are. So if our town wants a causeway, we'll probably have to build it ourselves. Either that, or sea level rise will eventually cut us off from the outside world.

I don't know if Chiara will still be living here by then, but this town will always be where she grew up. I know how precious my childhood memories are, and there are few things Chiara enjoys more than hearing me tell stories from when I was a boy on the farm. I live 2,900 miles away from that farm now, yet I still see its every detail in my mind. In her mind, Chiara too will always be able to revisit the place where she grew up. But at some point it may become impossible to visit here in person, at least by automobile. The only access may be by boat or on foot, just as it was when white people first settled here 160 years ago.

"What Made the Land So Salty?"

Sadia, the Bangladeshi mayor's daughter, is not immediately threatened by sea level rise: she lives in northern Bangladesh, 220 miles from the ocean. But earlier in my travels in her native land I met another girl about Chiara's age, named Uma, who lived in the far southwest, close to the Bay of Bengal. Nearly every Bangladeshi I encountered had brown eyes, but not Uma. Her eyes were the most extraordinary shade of green I've ever seen—the color of jade, yet somehow translucent. She was at the age when a toddler can stand but not yet walk, so her mother carried her around on her hip. When I arrived in their village and was introduced, I brought my palms together in front of my heart, smiled, and bowed my head—the customary local greeting. Looking up, I found myself staring into Uma's eyes and was instantly mesmerized. She broke the spell by bringing her tiny hands together and bowing her head in return, a grownup gesture that delighted her mother and melted my heart.

Bangladesh sits at the foot of the largest mountain chain in the world, the Himalayas, and 92 percent of the Himalayan snowmelt flows through the country on its way to the sea. The Ganges and the Brahmaputra are the largest carriers of this snowmelt, and although they rank among the mightiest rivers in the world, they are but two of fifty-four major tributaries that course through Bangladesh. In essence the entire country is a delta floodplain. Factor in the two hundred inches of rain that falls during the average monsoon season and it is easy to see why some call Bangladesh a nation of water. Even in a normal dry season, roughly a third of the country's area is covered by water; in rainy season, two-thirds can be covered. This liquid abundance and the rich soil deposited by the snow-fed rivers make it possible for farmers in some parts of Bangladesh to harvest an enviable three crops per year.

Farmers in Uma's village and elsewhere in southern Bangladesh are not so lucky. Because the land they occupy is flat and very low-lying, sea level rise is turning their soil and water salty.

It took ten hours of driving to reach Uma's village, and the sights along the way told the story. Driving west and then south from the capital, I spent the first hours of the journey gazing on a landscape that was almost monotonous in its fecundity. Even in dry season, the land was a bright green, thanks to densely planted fields of rice, the staple of Bangladesh since ancient times. But the closer we got to the ocean, the more often the brilliant green of the north was replaced by a dull brown. The culprit? Salt.

"The salinity of the soil here was four to five parts per thousand before 1970, but now it has risen to around twenty parts per thousand," said Mizan, an activist with Caritas, one of Bangladesh's leading NGOs, who was traveling with me. "Before this increase, the land was used to grow rice and other traditional crops, but now this kind of agriculture is no longer possible. The salinity is too high to allow favorable productivity."

"What made the land so salty?" I asked.

"This is because of sea level rise," explained Mizan, whose accent was so thick I read back all of his quotes to make sure I had understood him correctly. Citing the conclusions of a study recently completed by British and Bangladeshi scientists, he continued, "The sea is rising by about three millimeters a year. This sounds small. [It amounts to one foot in the course of one hundred years.] Nevertheless, it has an effect. As sea level rises, it pushes salty ocean water farther inland. This salty water mixes with the fresh water of the rivers and the salt settles in the soil. This is why salinity has increased to twenty parts per thousand over the last forty years. Farmers can still grow rice, but the yield is going down and down."

"Mizan is correct, but that's not the only reason for this problem," added Alim, another Caritas activist. "Some years ago, India built a dam across the Ganges and began keeping more of the river's flow for itself. This meant less flow into Bangladesh. In the past, the flow of the Ganges pushed against the pressure from the sea. Now that the flow of the Ganges is weaker, the seawater pushes farther inland."

Late in the afternoon, after turning off the paved road and bouncing down a dirt track for ten minutes, we arrived in Uma's village. Jelekhali was a cluster of about fifty thatch-and-bamboo huts nestled beneath shade trees and surrounded by rice fields that were that familiar lifeless brown. The next morning I took a stroll around the village and saw that about a dozen shoeless boys had commandeered part of the rice fields as a cricket pitch. When I got close enough, I saw how thin their limbs and frames were. I knew what starvation looks like from my travels in the Sudan years before, and these boys did not suffer that level of deprivation. But malnutrition did afflict 41 percent of the children in Bangladesh, these boys probably among them. "Their fathers are so much poor, these boys cannot eat sufficient food for their abdomens," Bikash Raptan, an activist with a local NGO, Sundipti, said in fractured English. "They eat rice and fish but not sufficient. This is a very saline area."

"Global warming has a taste in this village. It is the taste of salt," wrote Henry Chu, a staff writer for the *Los Angeles Times* who toured a nearby village around the same time I visited Jelekhali. Indeed, at a village meeting I attended after the cricket match, one woman described how her son recently had asked for a glass of water, and she had had to walk nearly a mile to the next village because the water in Jelekhali was so salty. The scarcity of fresh water not only leaves people thirsty, interjected Raptan, it fosters diseases such as dysentery, diarrhea, and jaundice. An elderly man talked about a nearby rice field that previously grew two crops per year but now managed only one because of increasing salinity, a problem he said had been worsening for the last ten to twelve years.

People were not the only ones suffering. Uma's village is located barely two miles from the Sundarbans, a dense crisscrossing of shallow channels and sandbars that separate the landmass of Bangladesh from the Bay of Bengal. A UNESCO World Heritage site, the Sundarbans is home to the largest mangrove forest on earth, not to mention one of the highest concentrations of tigers still living in the wild. But now the forest's trees had started dying "in a way nobody has seen before, from the top down," Justin Huggler reported in the *Independent* of London. Ainun Nishat, a biologist who was the senior adviser to the Bangladesh unit of the Union for the Conservation of Nature, told Huggler that scientists weren't sure yet, but it looked as if the trees were dying because the water at their roots was growing more salty because of sea level rise.

Thirty-one square miles of the Sundarbans had disappeared over the past thirty years, in part because of rising seas, according to a study by Sugata Hazra of Jadavpur University in Kolkata. As sea levels rise, the Sundarbans figures to keep disappearing. The loss of mangrove forests will also compound southern Bangladesh's vulnerability to climate change, for mangroves provide good protection against cyclones, a recurring threat in the region: a mangrove's expansive tangle of roots and branches weakens the force of a storm surge before it reaches inland. If the Sundarbans disappears, so will the Bengal tiger.

Sea level rise is often perceived as a problem of the future, but in Bangladesh the future is now. Although sea level has only just begun its inexorable ascent, coastal Bangladesh shows how rising seas can ruin land long before they inundate it. The majority of Bangladesh is a delta, so even limited amounts of sea level rise have an exaggerated effect. A rise of three feet would put 20 percent of Bangladesh underwater and create 30 million refugees, according to Atiq Rahman, the director of the Bangladesh Centre for Advanced Studies.

This realization has begun filtering down to the village level, along with an understanding of who is to blame. One afternoon while visiting Uma's village, I suddenly had the feeling of being watched. I turned around and almost bumped into the young man staring at me. Slightly built, with wisps of dark hair above his lips, he wanted to practice his English.

"Hello, sir," he said softly, as a friend nudged him forward. "Please, sir, I want to converse with you a little."

His name was Rajivit. He studied science at the local high school, about a mile from the village. He and his friends were "astonished" by me, he said. Except on television, they had never seen a person with white skin before.

"Please, sir, I would like to ask you about climate change. I have learned in school that carbon dioxide is collecting in the atmosphere and this is causing the earth to get hotter. Is it true?"

"Yes, that's what the scientists say," I replied.

He nodded. "And I have learned that the rich countries have put these gases into the atmosphere. Is it true?"

"Mostly," I said. "But now China and India are releasing many of these gases as well."

He nodded again. "I have learned that this CO_2 will make the ocean rise and cover the south of Bangladesh with water. This village too will be covered with water. Is it true?"

Looking into the young man's beseeching eyes, I hesitated to tell the truth but could not tell a lie. "I'm afraid that could happen someday, yes. The scientists aren't certain, but they believe it could happen."

"That is a big problem, sir," he replied. "Please, sir, how do we solve this problem?"

A Choice Between Pain and Disaster

Its residents often overlook the fact, but New York is a city of islands. Only one of its five boroughs, the Bronx, is part of the North American mainland. The city's other four boroughs are either islands outright, like Manhattan and Staten Island, or they are part of a larger island, as Brooklyn and Queens occupy the western edge of Long Island.

An island is inherently vulnerable to climate change. Along with delta areas such as Bangladesh and coastal zones such as California, the world's inhabited islands will be among the first targets of sea level rise, stronger storms, and other impacts of climate change. After all, there is only so much room on an island in which to take shelter from the sea. Vulnerability is especially high if the islands are low-lying and densely populated, as much of New York City is.

One morning in June 2007, I took a walk in lower Manhattan with Michael Oppenheimer, a New York native who happens to be one of the world's foremost climate scientists. In the late 1980s, as chief scientist of the Environmental Defense Fund, Oppenheimer had helped to convene meetings that influenced the creation of the IPCC. Since then, he had contributed to three of the four major IPCC reports, including the *Fourth Assessment Report* of 2007, and had become a professor of geosciences and international affairs at the Woodrow Wilson School of Princeton University. We met at the southern tip of Manhattan, near the spot where Dutch settlers had purchased the island in 1625 for the infamous sum of $24. Directly ahead of us stood the Statue of Liberty, its silhouette fuzzy against hazy silver skies. To our left, the view stretched to the edge of Brooklyn; to our right was the New Jersey coast.

"New York is surrounded by water, so three feet of sea level rise obviously could have a major impact here, especially if you factor in the surges you get from storms and hurricanes," Oppenheimer said. "During severe coastal storms, miles and miles of Brooklyn would be submerged. We can't see Queens from here, but the same would happen there. In Manhattan, flooding would be worst at this end of the island. Wall Street would be inundated. But you'd also get flooding from the Lower East Side all the way up to the United Nations in midtown, which is a very vulnerable area. LaGuardia and Kennedy airports would both be underwater. So would much of the subway system."

"How could New York function without airports?" I blurted out, unwittingly revealing my outsider's perspective. "Could they build seawalls to protect them?"

"How would New York function without subways?" Oppenheimer shot back, a grin spreading beneath his salt-and-pepper mustache. "New York would be nothing without mass transit. But yeah, they could build seawalls around the airports. You'd also have to raise the roads that lead to the airports, though, or what's the point? That would get very expensive."

Sea level rise, as Oppenheimer noted, also makes hurricanes and other ocean storms more dangerous. The dynamic is simple: the higher sea levels are, the more water a given storm can push inland. The sea around New York, the city government has reported, rose by nearly one foot over the course of the twentieth century. Even that relatively small increase, the government noted, has significantly increased the odds of a so-called 1-in-100-year flood. To be clear, this term does not refer to a flood so large that it happens only once every hundred years; rather, it means that there is a 1-in-100 chance *every* year that an extra-large flood will occur. As sea level rise continues, the government said, the annual probability of such extra-large floods striking New York City will increase from 1 in 100 to as much as 1 in 72.5 by the 2020s. By the 2050s, it will increase to 1 in 20.

This is not good news for a region that has long been visited by storms and even hurricanes. The worst storm in New York's history took place in 1938, when a Category 3 hurricane made landfall on Long Island, just east of the city. The storm killed more than 682 people and caused property losses estimated at $4.7 billion (in 2005 dollars), making it the seventh-costliest storm in U.S. history. Scientists have warned that the New York area is historically overdue for another major hurricane.

Nevertheless, it was the cultural implications of sea level rise that seemed to bother Oppenheimer as much as anything. "My great-grandfather came to New York in 1857," he said as we turned to walk uptown. "I've gone to see the buildings where his family lived and the graveyard where he is buried, near the border of Queens and Brooklyn. I believe people don't live only for themselves but as part of the flow of history. We have a duty to the future, to the people who will come after us, to leave them a livable planet. We also have a duty to the past, to the people who got us here, to preserve what they left behind."

After a few minutes of walking, we were staring down at a second potential loss: Ground Zero, the huge hole in the ground where the World Trade Center once stood. Surrounded by wire security fencing, the eight-square-block area was crawling with construction workers preparing to erect a memorial to the victims of the September 11 terrorist attacks. "Our society is building something here that we expect to last for hundreds of years, but this area will be very susceptible to flooding long before that," Oppenheimer said. "If we get the upper end of what the IPCC projected in the *Fourth Assessment Report*, about two feet of sea level rise by 2100, that means that every thirty years or so a flood will come along that is large enough to shut this area down for days. But if Stefan Rahmstorf's numbers are right [Rahmstorf was a physicist at the Potsdam Institute for Climate Impact Research in Germany and an IPCC contributor], and sea level rise turns out to be four and a half feet by 2100, then in this area such flooding would on average become a yearly occurrence.

"Everything depends on how fast this happens," Oppenheimer said as he paused to buy a pretzel from a street vendor. "A lot depends on the ice sheets in Greenland and Antarctica. If we were to lose the Greenland ice sheet completely, you're talking about twenty-three feet of sea level rise. That would put a lot more of Manhattan underwater and most of Brooklyn and Queens, too. Now, you have to realize that that much sea level rise would probably take hundreds of years. But you also have to realize that it would be an irreversible process."

Here Oppenheimer was touching on one of the most complicated and contentious issues in climate change science. Nothing in the IPCC's *Fourth Assessment Report* had generated more immediate controversy than its estimate of how far and fast sea levels would rise. To the astonishment of many, the report appeared to scale back previous projections of sea level rise, even though the bulk of the report said that climate change in general was getting worse. In 2001, the *Third Assessment* had said that sea levels could rise as much as 88 centimeters (34.5 inches) by 2100. The *Fourth Assessment Report* estimated a maximum rise of 59 centimeters (23 inches) by 2100. In other words, the worst-case scenario seemed to have shrunk from nearly three feet of sea level rise to just shy of two feet.

But a careful reading of the report reveals why the estimate decreased: the IPCC scientists had excluded the role of melting ice sheets from their calculations. They did this not because the ice sheets were not melting—media reports had *shown* them to be melting—but because the scientists did not understand the process well enough to predict its future course. The upshot was that the *Fourth Assessment Report*'s estimate amounted to an optimistic scenario, for it assumed that melting ice sheets would contribute zero additional sea level rise. This distinction was lost on the outside world, however. Being good scientists, the IPCC authors made sure to mention and explain their exclusion of the ice sheets. But the passage was buried deep in the report rather than highlighted in executive summaries, thereby assuring that the media overlooked it. Deniers of climate change were delighted; they seized on the lower estimate as proof that the dangers of climate change were exaggerated.

I asked Oppenheimer, who was centrally involved in the report, why the role of ice sheets was excluded.

"Because we know from direct observations of the current melting that our previous models of ice sheet behavior were wrong," he replied. "At this point, we simply don't know enough to make solid estimates of future melting. So we're waiting for better models and information."

"But the *Fourth Assessment* said that global temperatures are guaranteed to keep rising for decades," I said. "Doesn't that mean ice sheets *have* to keep melting?"

"Not necessarily," said Oppenheimer. "We know from history that ice sheet melting can start and stop, which implies that the current melting might also stop. So we need to find out what is *causing* the current melting. Is it global warming? Or is it something else?"

This struck me as good news—maybe the Greenland and Antarctic ice sheets weren't doomed after all. But it turned out that Oppenheimer was plenty worried about just that possibility. There was a good chance, he told me, that Greenland was "nearing the point where it melts away irreversibly, assuming warming is sustained." He found Antarctica even more troubling: the melting of its Amundsen Sea Embayment alone could add 1.5 meters (4.5 feet) of sea level rise, though it would take centuries for the process to play out. For now, he stressed, scientists simply did not know enough to determine how close the ice sheets were to the point of no return. With enough research, they might solve the mystery within the next twenty years, he estimated, adding, "At that point, if we saw that the ice sheets were melting fast enough to cause, say, ten feet of sea level rise over the coming three hundred years, city leaders would probably decide to build a seawall around New York, or at least the core parts of the city."

"You wouldn't advise the city to start considering a seawall now?"

"No, I wouldn't. Building a seawall is a very expensive thing to do, and I think we still have a fifty-fifty chance of avoiding the melting of the Greenland ice sheet. The uncertainty is over how much additional warming is required to cause that melting—is it 1°C or 4°C? We don't know yet. All we know at the moment is, we're flying blind. In which case the sensible thing to do is slow down."

We concluded our walk by heading west to a park along the Hudson River. The sun had broken through now; the grass was dotted with people—mothers with infants, office workers on lunch break—enjoying the day. Looking upriver, Oppenheimer pointed out a sports field where his eight-year-old son played baseball. Suddenly bells clanged behind us and a wave of students poured out of Stuyvesant High School, laughing and pushing as they ran past. I asked Oppenheimer if he expected those kids and his son to inherit a livable New York.

"We were foolish not to act sooner, but it's never too late to act," he responded firmly. "Global warming can always get worse. If it's too late to avoid a three-foot sea level rise, it may not be too late to avoid the collapse of the ice sheets. We still have a choice, even if it's only a choice between pain and disaster."

3 | My Daughter's Earth

We basically have three choices: mitigation, adaptation and suffering. We're already doing some of each and will do more of all three. The question is what the mix will be. The more mitigation we do, the less adaptation will be required, and the less suffering there will be.

—JOHN HOLDREN, science adviser to Barack Obama

ONE MONTH AFTER Barack Obama clinched the Democratic Party's 2008 presidential nomination, one of his top advisers held an invitation-only conference in Washington, DC. John Podesta had been former president Bill Clinton's White House chief of staff; in a few months, he would head Obama's transition team, helping the president-elect choose senior staff and Cabinet members. The conference Podesta organized was not the usual top-down Washington gathering where a few big shots give speeches, the audience listens dutifully, and everyone networks furiously during the coffee breaks. Podesta's conference, organized with colleagues from the Center for American Progress and the Center for a New American Security, was structured as a war game. Its subject was climate change.

Frequently employed by military and corporate leaders, war games test a given organization's readiness to cope with real-world crises before the crises happen. Participants are divided into teams. Organizers present the teams with plausible if extreme scenarios: a surprise nuclear attack on the United States, say, or the collapse of a fast-food company's supply chain. The teams then respond to the hypothetical crisis as best they can. Careful records of the proceedings are kept, for the goal of a war game is not so much to win as it is to draw lessons that can be applied back in the real world.

The focus of Podesta's war game was the military and humanitarian implications of climate change, and his insider clout helped him assemble an impressive collection of experts to explore the problem. Included among the approximately forty participants in the game were former senior U.S. military commanders, the editor of one of Japan's biggest newspapers, the chair of the Green Party of Germany, and the man who would become President Obama's chief international negotiator on climate issues, Washington lawyer Todd Stern. The star attraction, appearing live by satellite from India, was Rajendra Pachauri, the scientist who chaired the IPCC, which had been awarded the Nobel Peace Prize six months earlier.

The war game was set in the year 2015. Podesta, cast as the secretary-general of the United Nations, had just convened an emergency summit in New York. Addressing diplomats from the United States, the European Union, China, and India — the world's four leading climate powers — the secretary-general said he had brought them together to respond to a crisis that was in danger of spinning out of control. Global greenhouse gas emissions were still climbing, notwithstanding the IPCC's recommendation back in 2007 that they peak no later than 2015. What's more, the impacts of climate change were now becoming too obvious and costly to ignore.

A few months ago, the secretary-general reminded his audience, a Category 5 hurricane had scored a direct hit on Miami. He offered no details, but anyone who recalled what Hurricane Katrina had done to New Orleans could begin to imagine the devastation. But only begin: although the destruction Katrina visited on the Gulf coast was massive, Katrina had been "only" a Category 3 hurricane. The Category 5 storm that hypothetically hit Miami in 2015 was significantly more powerful. And, the secretary-general recalled, a second major American city,

Houston, had been hit by a Category 4 storm a few years earlier, in 2011.

Meanwhile, Podesta continued, Europe was suffering a fifth straight year of crippling drought in 2015. Terrible drought had also struck Africa. With crops failing, millions of Africans had fled their homes. Many were heading to Europe, where border tensions were rising.

Perhaps the region of greatest concern, however, was South Asia. Bangladesh had suffered a massive cyclone in 2013 that killed 200,000 people and left hundreds of thousands homeless. India's army had responded by closing the border between the two countries, but here, too, tensions were high; 250,000 Bangladeshi refugees were camped along the border. And India had its own troubles. Rice and wheat harvests had been ruined in 2014 when the monsoon rains came late. When the rains finally did fall, they caused extreme flooding in many cities. Partly as a result, commodity prices were now soaring the world over, spreading hunger and sparking food riots that in turn threatened political stability.

The secretary-general said he was especially concerned about the migrations all of this extreme weather had triggered. Calling the millions of people on the move "climate change refugees," he said they presented a huge humanitarian challenge. They also threatened international security, he said, a diplomat's way of warning that they could cause wars — for example, by crossing borders of nations that did not welcome them.

Suddenly, the year 2015 seemed very close — so close that I found myself hoping that the war game organizers had exaggerated the dangers for dramatic effect. Despite all I had learned while researching this book about the early arrival of climate change, I was shocked that such acute impacts could occur so soon. Checking the briefing book the organizers had distributed to the war game's participants, I read that the scenario Podesta had laid out was obviously an invention, created for the purposes of the game. But it was an invention based on the latest climate science, as contained in the IPCC's *Fourth Assessment Report* and vetted by scientists from the U.S. government's Oak Ridge National Laboratory and other leading institutions.

Still hoping that some exaggeration had taken place, I asked some of the scientists at the war game whether the specifics of the scenario—massive hurricanes demolishing Miami and Houston, millions of refugees in flight—were truly possible by 2015.

"Yes, I'm afraid it is possible," replied Kris Ebi, the public health expert who had been warning college kids about the end of pleasant summers. "Of course, no one can predict exactly what events will unfold as global temperatures continue to rise, and it's highly unlikely that this precise scenario will come to pass. But a scenario very much like it is entirely plausible. The science is pretty clear on that."

To be sure, the impacts hypothesized in Podesta's war game might not stem entirely from global warming. Like the extraordinary 2003 heat wave David King had instructed me about, extreme events often can be manifestations of the natural long-term variability of weather. But just as the rising baseline of temperatures caused by global warming had accounted for "about half" of the excessive heat of 2003, according to King, so would the rising baseline increase the odds of the extra-strong droughts and hurricanes envisioned in Podesta's war game.

What is undeniable is that my daughter's Earth—the planet Chiara inhabits over the next fifty years and beyond—will differ considerably from the planet her father and his generation occupied. In the previous chapter of this book, we examined what three feet of sea level rise could mean for the climate change generation. But sea level rise is only one of the impacts our children will encounter. If we are to prepare our societies for these impacts, we must learn as much as possible about what they will be. Granted, our knowledge will be imperfect, but that is better than no knowledge at all.

What Lies in Store

Toward that end, the following passage draws on the latest peer-reviewed studies to offer an overview of the most important impacts projected to emerge over the next fifty years. Supporting documentation can be found in the Notes section at the back of the book, but be advised: climate science has been progressing so fast that some of what I write here in early 2010 may be outdated before long. Interested readers are invited to consult my website, http://www.markhertsgaard.com, where I will post updates as warranted.

I should also emphasize that the following impacts are the minimum the scientific community expects. They reflect both the lag effect of past emissions and the time it will take to halt all global emissions (assumed here to be at least twenty years). Roughly speaking, these impacts correspond to a scenario in which average global temperatures rise no more than 2°C (3.6°F) above the level that pertained prior to the Industrial Revolution—that is, the level under which our civilization developed and to which the planet's ecosystems have adapted. Alas, this appears to be about the best future Chiara, Sadia, and the rest of the climate change generation can hope for. If humanity delays the shift to a low-carbon economy, the impacts will be even greater.

Harsher Heat Waves (and More Power Blackouts)

Begin by recalling David King's warning that the record summer heat of 2003 was a harbinger of summers to come: by 2050, Europeans will be experiencing summers as hot as 2003 one year out of two. The effects will range from minor personal concerns—greater discomfort, sweatier bodies, shorter tempers—to deadly serious ones. King told me the extra heat of 2003 had killed 31,000 people, according to European government figures—in other words, about half as many people as the U.S. war dead in Vietnam. But later in my reporting I learned that the 31,000 figure was a gross underestimate. A study sponsored by the European Union in 2008 reexamined the data and concluded that the 2003 heat wave caused at least 71,449 excess deaths.

Places that are hot today will be much hotter in the future. Residents of New York City now endure an average of fourteen days a year when the temperature is over 90°F. By the 2020s, the percentage of extremely hot and muggy days will roughly double; temperatures will exceed 90°F twenty-three to twenty-nine days a year. By the 2050s, the percentage will nearly triple, to twenty-nine to forty-five days a year. And instead of two extreme heat waves a year, New Yorkers in the 2050s will experience four to six of them.

Inland areas will experience significantly more temperature rise than coasts—not good news if you live in Chicago. Or Madrid or Paris. Or Delhi or Chongqing. In 1995, Chicago endured five days of record heat and humidity that left 739 people dead and countless others ill, lethargic, and plain exhausted. That kind of weather will occur much more frequently in future summers. By 2040, St. Louis will endure heat waves equivalent to 1995 Chicago's three times every summer. Historically, the state of Ohio has experienced three straight days of 95°F only once a decade. By 2050, this will occur in three summers out of four.

The suffering will be greatest among the poor, for whom air conditioning is usually an unattainable luxury. In India, the summer of 2003 was even hotter than in Europe, though India's plight got nowhere near the same attention from the world media. "In Paris, it was the isolated and the elderly who died from the heat, while in India it tended to be young male workers who had no choice but to keep laboring until they just fell over from heat exhaustion," said Richard Klein, an adaptation policy expert at the Stockholm Environment Institute. "The lesson is that adapting to climate change is not just a matter of technology. More air conditioning would have helped the elderly in Paris, but the young men in India needed different social arrangements, such as the right to take a break without losing their job."

A second casualty of higher temperatures may be the availability of electricity. In Europe, the 2003 heat wave depleted rivers, causing hydroelectric stations to produce less power. The heat also raised river water temperatures too high to cool nuclear reactors, causing power station shutdowns in France. A record heat wave in California in 2006 also triggered power blackouts as people used more air conditioning, increasing stress on transformers and power lines. Because nighttime temperatures also stayed high, the transformers and lines had less opportunity to cool off and therefore failed more often. The result was blackouts that left 2 million Californians without electricity at a time when air conditioning was literally a lifesaver.

Air conditioning is but one of the electrical necessities of modern society; it is scarcely an exaggeration to say that today's information society is addicted to electricity. Imagine life without the countless other devices that run on electricity: computers, cell phones, ATM machines, refrigerators. Unless steps are taken to bolster the reliability of the electricity supply, global warming could make blackouts a regular feature of future summers, bringing social havoc and immense human suffering.

Stronger Storms, More Disasters

Another reason to expect more blackouts is that hurricanes and other weather-related disasters are projected to increase in intensity as temperatures rise. People who reside on or near America's Gulf and Atlantic coasts already know that hurricanes can destroy their communities overnight, as do their counterparts in the Caribbean, along the coasts of Japan, China, southern Asia, and much of the South Pacific. But the likelihood of such mega-storms, and mega-damage, will increase significantly in the years ahead.

The IPCC's *Fourth Assessment Report* projected that Category 4 and 5 hurricanes (the most powerful on the scale, capable of destroying entire cities) will become more frequent as warming continues, though the overall frequency of tropical storms may diminish. Currently, there is an average of thirteen such mega-storms a year worldwide, though not all of them make landfall. However, the 1°C of future temperature rise that is locked in will increase that number by 31 percent by 2050, according to a study published in *Nature* in 2008. That means at least four more mega storms will occur each and every year by 2050, raising the odds that population centers will be hit.

Major disasters have already been trending upward for nearly two decades, according to data collected by the Munich Reinsurance Company, one of the world's leading reinsurance companies. (Reinsurance companies insure retail insurance companies; they were the first part of the business community to sound the alarm on climate change, in the mid-1990s.) John Holmes, the UN's coordinator of emergency disaster relief, reported that fourteen of the fifteen major relief operations that his team mounted in 2007 were in response to floods, storms, and other climate-related events. In 2008, nine out of ten major disasters were weather-related, causing up to $200 billion of damage. Yet neither governments, businesses, nor citizens were heeding the warnings, said Holmes, who added, "The risks of mega-disasters in some . . . mega-cities is rising all the time." The humanitarian organization Oxfam has projected that extreme weather could affect 375 million people a year by 2015, and the international relief system "could be overwhelmed."

As always with climate change, it is the poor and vulnerable who figure to be hurt most. The human suffering and social havoc that engulfed New Orleans in the wake of Hurricane Katrina show what can happen when a community's defenses against mega-storms are inadequate, which many are, especially in poor countries. The International Committee for the Red Cross has warned that stronger storms threaten to reverse decades of progress in the fight against poverty. "The kind of devastation caused by Hurricane Mitch that hit Central America in 1998 or the 2004 floods in Bangladesh and India shows that . . . many of the slow, hard-won gains in human development of the last few decades . . . could be swept away in a matter of hours," said *Up in Smoke,* a report the ICRC coproduced with Oxfam, Greenpeace, the New Economics Foundation, and other environmental and humanitarian organizations. Noting that it costs seven times more to cope with a disaster after the fact than to install safeguards in advance, *Up in Smoke* argued that climate change adaptation must become an explicit part of the fight against poverty, explaining, "Lifting people out of poverty is the best way to reduce the number who have to be lifted out of the mud, floodwaters or drought when disaster strikes."

More Disease and Pestilence

Climate change is the number-one threat to global public health in the twenty-first century, according to the *Lancet,* the world's leading medical journal. Already, climate change kills 150,000 people a year, according to an estimate by the World Health Organization cited in the *Fourth Assessment Report,* though again this figure appears to underestimate the true death toll. It turns out that the 150,000 estimate was made in 2002 and took into account only deaths from malnutrition, malaria, and diarrhea caused by contaminated water (a common result of floods), says Paul Epstein, the associate director of the Center for Health and the Global Environment at Harvard Medical School. The calculation excluded the effects of heat waves, crop losses due to an expected increase in pests, and a range of deadly diseases that the WHO itself has since predicted are bound to increase as global warming intensifies. Indeed, the WHO has already identified more than thirty new or resurgent diseases during the past thirty years—the most since the Industrial

Revolution—and attributes the problem to ticks and other pests expanding their ranges to areas that are unprepared for them. For example, mosquitoes that spread malaria have been found in such previously inaccessible areas as the highlands of Papua New Guinea. The range of asthma, a disease approaching epidemic levels in some U.S. inner cities, will expand in the years ahead, because hotter weather stimulates pollen and fungal production and increases ground-level ozone. Even a 1°C rise in temperatures—highly likely over the next fifty years—could bring back the infamous bubonic plague, the disease responsible for the Black Death of the Middle Ages that claimed an estimated 20 million lives.

Less Water (Except When There Is More)

When I think of Chiara's future under climate change, nothing worries me as much as water. California residents often forget that their state is a desert; the next fifty years will remind them. Researchers led by Columbia University's Richard Seager reported in *Science* that record drought *will become the norm* (my emphasis) across the western third of North America, including much of California, by 2050. So, good luck, Los Angeles, San Diego, and California's farmers; you're going to need it.

Water supplies are already extremely stressed today in most of the southwest of the United States, as suburban sprawl, population growth, rampant waste, and rising consumption rates confront static or declining supplies. Lake Mead, which is the largest reservoir in the United States by volume and supplies water to Las Vegas, Phoenix, and other major cities, will be unable to meet all customers' needs 88 percent of the time by 2050 because of declining runoff from mountain glaciers and snowpacks. The Colorado River as a whole, which supplies water to 27 million people across the Southwest, will not supply the total volume allocated to customers 60 to 90 percent of the time by 2050. With cities and farmers already squabbling with neighbors today over tight water supplies, the "water wars" of the West seem destined to intensify in the years ahead.

Overseas, a similar fate will befall much of the Mediterranean basin, Australia, Brazil, and southern Africa, each of which has suffered record drought in the past few years. Yet even as drought becomes more common, so will record rainfalls, raising the likelihood of dangerous flooding that could cause dams to fail. If a huge surge of water down the Yangtze River in China were to overwhelm the Three Gorges Dam, tens of millions of people could die, Chinese officials told a Western expert I interviewed, which would make it perhaps the deadliest disaster in human history. (The expert asked not to be named because his hosts would resent him sharing such sensitive information.)

It is a paradox that global warming will cause both deeper droughts and fiercer floods. As global warming intensifies, more rain will fall in intense bursts, followed by longer periods when there is no rain at all. Meanwhile, higher temperatures will cause rain that *has* fallen to evaporate more rapidly; thus soils will dry more quickly. Hence at various times there will be both less and more water than usual. Australia's recent history provides a preview. In 2007, drought became so severe that every major city in Australia had emergency water restrictions in place. Food production fell 25 percent. The incumbent government—which had resolutely denied that global warming was a problem—was voted out of office. Two years later, though, the problem was flooding. Massive deluges in eastern Australia closed highways, isolating some twenty thousand people from food and medical supplies. Untold numbers of livestock were killed, and residents of New South Wales were warned to be on the lookout for venomous snakes that had been spotted in some town centers.

Perhaps the greatest water-related hazard is that rising temperatures will melt snowpacks throughout the world. In California, snowmelt is the source of nearly one-third of the freshwater supply. (The snowpack atop the Sierra Nevada, which run along California's eastern border, provides roughly a quarter of the total supply, and the snowpack of the Rocky Mountains far to the east feeds the Colorado River, which supplies about 15 percent of the total.) Higher temperatures will shrink the Sierra Nevada snowpack by 25 to 40 percent by 2050, according to the California Department of Water Resources. Specifically, rising temperatures will cause precipitation to fall as rain rather than snow more often; hotter weather will also cause snow that does fall to melt and run off down the mountains sooner and faster. That may or may not affect the total amount of water involved, but it will certainly affect the timing of the runoff, which will have much the same effect. More water will run off during late winter and spring, leaving that much less at our disposal in summer, the very time crops and humans need it most.

Thus, a state whose water supply is already overallocated—a bureaucrat's term, meaning that the demand on the part of cities, farmers, commercial interests, and environmental requirements already exceeds the water available—and whose population is expected to grow will face harsh additional constraints on the water available to it. Much the same challenge faces Seattle and other municipalities in the Pacific Northwest, where the snowpack atop the Cascade Range is likewise projected to dwindle.

Again the trends are equally worrisome internationally. In Asia, at least 500 million people obtain some of their drinking and irrigation water from the 42,298 glaciers atop the Himalayan mountains, a mass of snow and ice long known as the "Third Pole" of this planet. NASA's Hansen has warned that "continued 'business-as-usual' emission of greenhouse gases and black soot will result in the loss of most Himalayan glaciers this century. . . ." His colleague Yao Tandong estimates that 40 percent of the glaciers will disappear by 2050. In Asia, an estimated 500 million people obtain some of their drinking and irrigation water from the Himalayan snowpack, which scientists warn could largely disappear by 2050. In Africa, Mark Lynas reports in *Six Degrees*, one of the essential books on climate change, the glaciers atop the Rwenzori Mountains in Rwanda "are expected to be gone within the next two decades." (Lynas's title refers, of course, to 6°C, or 10.8°F.) In South America, scientists expect most Andean glaciers to disappear by 2030. The situation is especially dire in Ecuador, where Quito, the capital, depends on glaciers for half of its water supply.

Drought is the overriding danger as climate change intensifies—"the elephant in the room," to quote climate historian Brian Fagan. Floods may attract more media coverage, but historically droughts have killed far more people. "Floods kill thousands, drought can kill millions," goes the adage recited by one expert. Drought is especially punishing for the hundreds of millions of subsistence farmers around the world for whom rain is the only source of water. By roughly 2025, the number of people living in water-stressed countries will increase from 800 million to 3 billion—an amount equal to nearly half of the current global population.

Less Food, More Fires

Combine rising temperatures with shrinking water supplies and the implications for food production are obvious. The U.S. Midwest, sometimes called the world's breadbasket, will face severe stress. "The temperature rise that is already in the pipeline for the next thirty years is projected to reduce corn yields in the Midwest by 10 to 20 or even 30 percent [if no adaptation measures are taken]," said David Lobell, a professor at Stanford University's Program on Food Security and the Environment. "In the southeast of the U.S., corn yields could fall by 50 percent." And those figures are likely to be underestimates, Lobell added, for they are based only on the influence of higher temperatures; climate change will also likely reduce water supplies, further stressing production. Of course, higher temperatures and drought often go together, causing more concern. Lynas points out that in the Middle Ages, a mere 1°C of temperature rise caused the prairies of Nebraska and other midwestern states to dry up and revert back to the sand dunes that had covered the area thousands of years earlier. If that happens again, Lynas speculates, we could see a replay of the Dust Bowl disaster of the 1930s, as some of the richest soil in the world takes flight and blows away, carrying with it the source of food for millions of people around the world.

China, the world's most populous nation, home to one of every five humans on the planet, is also facing an extreme challenge to its food security. The Chinese government projected in 2007 that the nation's yields of wheat, corn, and rice could decrease by 37 percent in the next few decades if no adaptation measures are taken. Of course, China could compensate by buying more food on the world market, but that would push prices up internationally, raising the incidence of hunger, disease, death, and political instability. Recall how soaring food prices in 2008 led to street riots in over a dozen countries, including Bangladesh, the Philippines, Indonesia, Egypt, and Haiti. Such tumult is all the more likely considering that climate change will be undermining food production in many if not most poor countries. In South Asia, for example, the area suitable for growing wheat will fall by 50 percent by 2050. Africa will be hit particularly hard, with large declines in rainfall projected in eastern and southern Africa and across the Sahel.

Hot, dry weather bakes soil but burns foliage, raising the probability of wildfires. During the summer of 2003, forest fires in Portugal caused $1.5 billion in damage, Lynas reports, adding that over the next few decades "two to six weeks of additional fire risk can be expected in all countries around the Mediterranean rim." In the American West, the weather of the last twenty years sparked four times as many large fires as during the previous twenty years. Firefighters are worried: in 2006, their Association for Fire Ecology warned, "Under future drought and high heat scenarios, fires may become larger more quickly and be more difficult to manage." In 2009, a study by Harvard University scientists endorsed the firefighters' concerns. The area burned by wildfires in the American West could increase by 50 percent by 2050, the study found; in the Pacific Northwest and the Rockies, the increase could be as much as 175 percent.

Mass Extinctions

As temperatures rise over the next fifty years, the impacts on plants, animals, and the natural systems that make our lives possible on earth will be profound. Coral reefs, for example, provide coastal communities with vital protection against storms, as their underwater bulk breaks the force of incoming storm surges. Coral reefs are also a key foundation of the marine food chain, nurturing the fisheries that provide 1 billion people on earth with their primary source of protein. But coral reefs are effectively doomed by the inertia of the climate system. Higher temperatures serve to bleach and kill coral; some scientists expect 98 percent of the world's coral to be gone by 2050.

Coral also suffer grievously from the acidification of the ocean caused by greenhouse gas emissions. As oceans absorb more and more of the carbon dioxide spewing from humanity's power plants, vehicle tailpipes, and livestock operations, seawater has grown more acidic than it has been anytime in the last 800,000 years. High acid levels not only degrade coral; they make it very difficult for shellfish to make thick enough shells to survive.

The living dead is the term scientists have coined for the existing specimens of coral and other species that seem bound to go extinct, such as polar bears. In 2007, the Arctic experienced a record amount of summer ice melt. "At this rate, the Arctic Ocean could be nearly ice-free at the end of summer 2012, much faster than previous predictions," NASA scientist Jay Zwally told the Associated Press. That scenario bodes ill for polar bears, who need ice to hunt; the U.S. Geological Survey has estimated that two-thirds of the world's polar bears will be gone by 2050.

I find it terribly sad that Chiara and the rest of the climate change generation will grow up in a world where polar bears and so many plants and animals are being pushed off the planet. And their extinction threatens our own, in two major ways. For the 1.5 billion people worldwide who are officially classified as poor, forests, oceans, and other ecosystems are a leading source of food, clothing, medicine, and other essential material goods. Scientists calculate that ecosystems account for 40 to 50 percent of the poor's economic consumption, suggesting that the loss of these ecosystems would halve the poor's already meager living standards. Beyond that, the ecosystems with which we humans share the planet provide "ecosystem services" that make an indirect but absolutely indispensable contribution to our species' survival. Humans often forget that we rely on plant, animal, and microbial species to maintain healthy soil, clean water, breathable air, and other necessities of a livable planet. As naturalist E. O. Wilson has observed, "We need [ants] to survive, but they don't need us at all." Without ants, earthworms, and other unsung creatures to ventilate it, the earth's topsoil would soon rot, ending food production. Without vibrant forests, water supplies would shrink. To study nature is to realize, to quote the old environmental axiom, that everything is connected. What we do to the polar bears, we do to ourselves.

"I Was Called a Traitor"

There is not a government in the world that is prepared for the impacts climate change will unleash. The United States lags notably behind, both at the federal and at the state and local levels, though there are exceptions: the state government of California and, as we'll see in the next chapter, the municipal governments of New York City, Chicago, and King County, Washington. But no one else comes close to the leaders in the field, the national governments in Britain and the Netherlands. The Dutch are developing a two-hundred-year plan to adapt their nation to climate change (no, two hundred is not a typo). They are spending approximately $1 billion a year to implement this plan, and they are making significant, sometimes controversial, changes in their society, as Chapter 5 of this book describes.

As for the British, the national government established an adaptation agency, the United Kingdom Climate Impacts Programme (UKCIP), way back in 1997. In 2002, UKCIP collaborated with the City of London on a report outlining the impacts climate change would have on the capital, from harsher summers and reduced water supplies to increased flooding risk. Since then, UKCIP has worked with scores of local governments and private businesses in Britain to help them prepare for coming impacts. In 2009, the British government became the first in the world to provide local maps of likely impacts of climate change, an invaluable resource for individuals, governments, and corporations that wish to prepare themselves. "We also announced that every UK government department has to prepare by 2010 both an adaptation and a mitigation plan looking decades ahead," said Robin Mortimer, the director of climate change adaptation at the Department for Environment, Food and Rural Affairs. "All government investments larger than 20 million pounds must be appraised for their climate risks as well."

The most visible example of the British commitment to adaptation is the Thames Barrier, a set of beautiful silver floodgates that stretch across the namesake waterway about eleven miles downriver from central London. When the barrier became operational in 1982, thirty years after the massive flood that motivated its construction, planners expected that it might have to close once or twice a year to keep ocean storm surges from inundating London. In the past decade, however, the barrier has been closing an average of ten times a year. "The barrier was initially designed to offer a 1-in-2,000-years level of protection," said Chris West, the director of UKCIP. "But sea level rise is projected to reduce that to a 1-in-1,000-years level by 2030." In response, the British government is prepared to add twelve inches of protection on top of the existing floodgates—a contingency built into its design—and to keep extending the barrier as necessary.

Despite these and other initiatives, Dutch and British adaptation experts are quick to acknowledge that they still have far to go. Adaptation remains the "poor relation" in Britain's climate strategy, West told me. "The government's mitigation programs, like the Carbon Trust and the Energy Savings Trust, well, I wish them luck, but they get fifty times more money than we do." Klein of the Stockholm Environment Institute, a Dutchman by birth, said, "It's probably true that the Netherlands is one of the global leaders in adaptation, but that only shows that you don't have to have done very much to count as a leader. The Dutch government is doing what it's doing because it has to. It is required by law to protect people from water, because no individual living six meters below sea level can do that for himself."

Adaptation emerged as a major bone of contention between poor and rich nations in 2001, at UN-sponsored climate negotiations in Morocco. Four years after the signing of the Kyoto Protocol, global greenhouse gas emissions were still increasing, and the governments of poor nations were growing increasingly concerned that they could be overwhelmed by a problem not of their making. "It was the back-to-back floods in 2000 and 2001 in Mozambique that got people's attention," recalled Youssef Nassef, who was Egypt's representative to the UN Framework Convention on Climate Change at the time. "To have bad flooding in a single year was nothing out of the ordinary. But two years in a row? Before then, climate change was seen as more of a problem for the future. You have to realize, policymakers in developing nations face many problems that are killing people today: poverty, disease, HIV-AIDS. But when floods hit two years in a row, we knew that climate change could no longer be overlooked."

The idea of climate change adaptation did not enter mainstream discourse until 2009, when developing nations' demand for adaptation assistance and funding from rich nations became a major stumbling block at the Copenhagen climate summit, as we'll discuss further in Chapter 10. But until that time, most governments of rich nations had barely begun to discuss adaptation, and most businesses, individuals, and civic institutions around the world had not even heard of the concept. Not until the IPCC's *Fourth Assessment Report* made it clear in 2007 that there would be major climate impacts, no matter what, did the European Union begin discussing adaptation, along with the governments

of Germany, Italy, and Spain. Japan, too, has lacked a formal adaptation policy, though this shortcoming is somewhat offset by the disaster preparedness procedures the country has developed to cope with its historic vulnerability to earthquakes, cyclones, and flash flooding. The impacts projected under climate change, however, have apparently not been integrated into this regime. In Tokyo, flood control depends on the G-Cans Project, a massive underground system that can pump two hundred tons of water per second out of rivers and into the harbor. But former senior city officials told me this system had reached its capacity and needed to be upgraded, a slow and costly process.

In the United States, public discussion of global warming in general has long lagged years behind the rest of the world, and the topic of adaptation is no exception. "You can't adapt to a problem you don't admit exists," Klein notes dryly. The Bush administration, which repeatedly downplayed the climate issue during its eight years in office, killed a key adaptation tool: the National Climate Assessment, a detailed analysis of the vulnerabilities of various regions of the United States and the possibilities for coping with them. The Obama administration reinstated the program and issued a 188-page report in June 2009. The report's projections mirror many of those summarized earlier in this chapter, such as an increase in the intensity of hurricanes and a rise in sea levels of three to four feet by 2100. Breaking from previous federal policy, the report emphasized that both adaptation and mitigation would be required to cope with climate change. Adaptation would be especially challenging, the report explained, "because society won't be adapting to a new steady-state but rather to a rapidly moving target. Climate will be continually changing, moving at a relatively rapid rate, outside the range to which society has adapted in the past."

By the time Obama entered the White House, many poor decisions had already been made in the United States, including in places that should have known better. In California, the officials building a new bridge across San Francisco Bay apparently didn't get the memo on climate change. The old bridge had been damaged by the earthquake of 1989. Planners made sure its replacement, a $6.3 billion investment, could withstand future quakes but did not bother to factor sea level rise into their calculations. "The entrance ramps to the Bay Bridge on the East Bay side are at sea level," said Gleick of the Pacific Institute. "Fifty years from now, if not sooner, those ramps will have to be raised. The intake structures of lots of power plants will have to be changed, too. There will be a lot of 'rip it up and replace it' costs to infrastructure in the years ahead."

Progress on adaptation has been slow partly because *adaptation*, until recently, was something of a dirty word in climate circles. Many activists charged that even to discuss adaptation played into the hands of the George W. Bush administration and others who resisted reducing emissions. "Until about five years ago, we were seen as part of the problem more than as part of the solution," West of UKCIP told me in 2007. Recalling a climate change conference in India in 2002, he added, "I remember an activist telling me, quite angrily, 'You're worse than irrelevant, because you distract people from what really needs doing,' by which he meant mitigation."

Most longtime proponents of adaptation can tell similar stories. "When I was advising the German government in the mid-1990s, I was called a traitor for suggesting that we needed to talk about adaptation, even though I also advocated an ambitious mitigation policy," said Hans Joachim Schellnhuber, the director of the Potsdam Institute for Climate Impact Research in Germany. Schellnhuber survived the accusation well enough—he went on to become German chancellor Angela Merkel's chief adviser on climate change—but he cited the incident as an example of how ingrained the bias against adaptation was, especially among those most determined to fight climate change. "The sense was that if you talked about adaptation, you were giving up on mitigation and playing into the hands of those who didn't want to confront the problem," says Schellnhuber, "though that was not how I looked at it."

Adaptation critics were not being paranoid. Beginning in the early 1990s, government officials, corporate spokespersons, and academic critics, especially in the United States, did invoke adaptation to fend off calls for mitigation. This camp argued that if climate change eventually turned out to be a real problem (a possibility they disparaged), the world could respond by adapting. Relying on adaptation, they said, would be less economically disruptive than cutting the consumption of oil, coal, and the other carbon-based fuels that are the lifeblood of the modern world.

Most governments and industries abandoned this line of argument as the scientific evidence for climate change solidified throughout the 1990s. But it remained the policy of the Bush administration at least through 2006, judging from what President Bush's science adviser told me in an interview. Defending Bush's rejections of mandatory emissions reductions, John Marburger said, "There is no question that mitigating the impact of climate change as it takes place will be much less [expensive] than the costs of reducing oil and coal use." Marburger got his terminology wrong—he talked about mitigating impacts when he meant adapting to them—but his meaning was clear: adaptation can hold the line until societies make the technological breakthroughs needed to solve the problem.

On the opposite side of the debate, even some environmentalists who intellectually recognize the case for adaptation continue to argue that it is politically dangerous to champion it. They fear that highlighting adaptation will undermine public and governmental support for mitigation. They also worry that any funding allocated to adaptation would mean that much less money for mitigation. Adaptation experts counter that such concerns are overblown. "Most adaptation initiatives will be undertaken by completely different agencies—for example, public health services, coastal protection agencies, agricultural research and extension services—from the agencies that should take the lead on mitigation, so you're really talking about separate budgets," explained Klein of the Stockholm Environment Institute.

Despite the resistance, adaptation seems bound to command more attention as the impacts of climate change become increasingly obvious and disruptive. Events on the ground have already made believers out of many key actors, including Al Gore (who cautioned against adaptation as late as 2008), UN Secretary-General Ban, governments of developing countries, the World Bank, the UN Development Programme, the Rockefeller Foundation, and the humanitarian and environmental NGOs behind the *Up in Smoke* report. And at the Copenhagen climate summit, adaptation was recognized as an imperative by virtually all parties to the negotiations. By no means did the summit resolve the issue, but there was broad agreement that rich industrialized countries have an obligation to provide substantial adaptation funding. U.S. secretary of state Hillary Clinton proposed transfer payments of $100 billion a year by 2020, though these were intended to help poor and vulnerable countries reduce their emissions as well as adapt to the impacts that are in store.

In the end adaptation is a necessary but insufficient response to climate change. "Adaptation by itself is not a solution for the rich, much less for the poor," said Saleemul Huq. "Unfortunately, the inertia of the climate system means that some impacts are locked in over the next thirty years. These impacts may not be catastrophic for everyone, but certainly poor people will suffer. The only way to limit that suffering is with adaptation. But if we don't have big cuts in emissions in the next twenty years, we'll have a global catastrophe even the rich can't manage."

Is 2°C Really a Safe Target?

The Dutch generally display a can-do attitude in the face of climate change, but even they warn that continuing with business as usual is not an option. "We participated in a study a few years ago called the Atlantis Project," recalls Laurens Bouwer, a researcher at the Institute for Environmental Studies in Amsterdam. "The question we examined was: If there were five meters [seventeen feet] of sea level rise by the year 2150, could we still protect the cities that lie within the deltas of the Thames, Rhone, and Netherlands river systems [in other words, cities such as London, Rotterdam, and Amsterdam]? Most of the experts we interviewed said no, we would have to abandon the country. Technically, our societies could make the changes needed, such as stronger sea defenses, and we could even pay for these changes. But the social and political tensions that would follow would be insupportable: investors wouldn't trust they could make their money back, people would be forced to live too close together — it just wouldn't work."

If current emissions trends persist, the concentration of greenhouse gases in the atmosphere will exceed 700 parts per million (ppm) by 2100, according to the *Fourth Assessment Report*. The temperature will increase by a minimum of 5°C (9°F) over the average temperature of the preindustrial era. The earth would be hotter than at any time in the past 50 million years.

It is questionable whether civilization could survive such temperatures and the impacts they would produce. In *Six Degrees*, Lynas listed some of the most dramatic impacts: The Greenland ice sheet would disappear, along with much of the Antarctic ice sheet, raising sea levels at least twenty-three feet, though again this process would likely take at least a century to complete. Much sooner, heat waves like that of 2003 would become annual events in Europe, while close to the equator vast swaths of Africa, Asia, and South America would simply become too hot to inhabit. Hundreds of millions would therefore face a choice: perish or flee (but flee to where?). Most sea life would be dead, thanks to the excess of carbon raising acidity levels in the ocean. As much as 90 percent of species could go extinct, raising the odds that *Homo sapiens sapiens* might vanish as well.

Alarmed by such prospects, many participants in the climate debate have proposed limiting the global temperature increase to 2°C (3.6°F) above the preindustrial level—that is, 2°C above the level in which our civilization developed and to which the earth's ecosystems have adapted. The 2°C target was first mentioned in 1990 by an advisory group to the World Meteorological Organization; it was reiterated by the IPCC in its *Third Assessment* (2001) and its *Fourth Assessment* (2007) reports. Later, it was endorsed by a wide range of governments—including the European Union and, in 2009, the Group of Eight rich industrial nations—as well as by many humanitarian and environmental groups. Because journalists rely on these entities for information, most news reports also end up implicitly endorsing the 2°C target.

The supporting argument, in brief, is that the projected impacts of climate change increase substantially once average global temperatures rise beyond 2°C above preindustrial levels. Of course, even 2°C will have effects; the impacts summarized earlier in this chapter are based on roughly 2°C of warming. But above 2°C, the argument goes, the impacts enter an especially dangerous realm: not only are they more damaging in themselves, but they are more likely to trigger "positive feedbacks" that will cause still more global warming and thereby perhaps give climate change an unstoppable momentum.

Positive feedbacks are self-reinforcing processes. For example, when rising temperatures cause polar ice sheets to melt, the white ice is replaced by dark water. While sunlight that hits white ice is largely reflected back into space, sunlight that hits dark water is largely absorbed. That absorption further warms the water, which further melts the ice, which warms more water, and so on—a positive feedback loop. Likewise, the permafrost that covers millions of square miles in Russia and Canada is more likely to thaw as temperatures rise. This is dangerous because permafrost contains vast quantities of carbon. The estimated 1,500 billion tons of carbon stored in the world's permafrost "is equivalent to twice the current amount of carbon dioxide in the world's atmosphere," said Pep Canadell of Australia's national science agency, CSIRO. When permafrost melts, this carbon is transformed into carbon dioxide or methane and escapes into the atmosphere. There, in another positive feedback, these greenhouse gases drive additional warming, which melts yet more permafrost and ice sheets and so on down a very dark path.

Positive feedbacks in turn can trigger the kind of abrupt, irreversible climate changes that scientists call "nonlinear." Hurricane Katrina, which strengthened from a Category 1 storm into a Category 5 after it encountered super-hot surface waters in the Gulf of Mexico, provided a sobering preview of how nonlinearity works. "Hurricanes are the mother of all nonlinear events, because small changes in initial conditions can lead to enormous changes in outcomes," said Schellnhuber, the German government adviser. "A few percent increase in a hurricane's wind speed can double its destructiveness under certain circumstances."

Although scientists apply the term *climate change* to all of these phenomena, Schellnhuber told me that *climate chaos* better conveys the abrupt, interconnected, wide-ranging consequences that lie in store. "It's a very appropriate term for the layperson," said Schellnhuber, whose early work as a physicist happened to focus on chaos theory, "for it refers to perhaps the most important concept that laypeople should understand about climate change—its nonlinear nature. I keep telling politicians that I'm not so concerned about a gradual climate change that may force farmers in Great Britain to plant different crops. I'm worried about triggering positive feedbacks that, in the worst case, could kick off some type of runaway greenhouse dynamics."

Unfortunately, there is ample precedent for this kind of abrupt shift into climate chaos. Although the human mind tends to think in gradual, linear terms, ice records and other historical data show that climate shifts, when they occur, tend to happen suddenly and exponentially. "Society may be lulled into a false sense of security by smooth projections of global change," Schellnhuber and his colleagues warned in a 2008 study. The study found that, if current emissions trajectories continue, a variety of abrupt changes could be triggered by the end of this century, including a substantial melting of the Greenland and Antarctic ice sheets, a die-off of the Amazon rainforest, and the collapse of the Gulf Stream, which accounts for Europe's moderate climate.

The urgency of reversing global warming is such that even champions of adaptation say it must be the top priority. "Environmentalists have tended to see adaptation as a capitulation and a distraction from mitigation, which is why it's been hard to get through to them about the importance of adaptation," said Suzanne Moser, the American adaptation scientist. "But if you start thinking about adaptation, you realize pretty quickly that we simply *can't* adapt past a certain point. If we let global warming get out of hand, our current outlook on adaptation will look quaint. So the major effort has to be thinking outside the box about how to reduce emissions. If I had $2 to spend on climate change, I'd spend $2.50 of it on mitigation. That's how important it is to keep global warming from getting out of hand."

But what does it mean, exactly, to keep global warming from getting out of hand?

Although many governments have endorsed it, the 2°C temperature limit is by no means truly safe. An increase to 2°C above preindustrial levels would cause sea level to rise twenty feet higher and possibly much more, though this increase would probably take place over at least two hundred years. "The only direct evidence we have for what happens at 2°C is from the last interglacial [period of earth's history, about 125,000 years ago]," said Michael Oppenheimer. "Sea level then reached six to nine meters, or twenty to thirty feet, higher than today." Oppenheimer added that 2°C of temperature rise would also cause "grievous damage and elimination of most of the [earth's coral] reefs, with large loss of biodiversity," as well as "substantial shrinkage of the Himalayan glaciers." Schellnhuber projected even greater levels of sea level rise from 2°C but added, "What really makes for sleepless nights is that you cannot exclude that [2°C] will trigger . . . runaway global warming." The latter concern was echoed by George Woodwell, the chief scientist of the Woods Hole Research Center. "You have to realize that a 2°C rise globally will mean a 6°C rise near the poles," said Woodwell; this, largely because of the aforementioned feedbacks whereby melting ice leaves behind dark water that absorbs more sunlight. "That would devastate boreal forests and thaw tundra, which would release large amounts of carbon dioxide and unleash dangerous positive feedbacks."

A 2°C rise globally would also mean at least a 3°C increase for much of Africa (again, because temperatures will rise farther in the middle of landmasses), devastating the continent's agriculture. A 2°C increase would also cause low-lying island nations to disappear beneath rising sea levels. These concerns help explain why more than one hundred governments representing a majority of the world's nations endorsed limiting global temperature rise to 1.5°C at the 2009 Copenhagen climate summit. "Anything more than that and we've had it," President Mohamed Nasheed of the Indian Ocean island state Maldives told me.

Of course, the safest course would be to have zero warming above preindustrial levels, but that is no longer possible. Humanity's previous emissions have already raised temperatures by 0.8°C since 1900, and there is an additional 0.6°C of warming in the pipeline because of the climate system's inertia, according to the IPCC. Adding this 0.6°C to the 0.8°C of warming already registered makes a total of 1.4°C that is already locked in. Therefore, to meet the 1.5°C target would require limiting additional warming to 0.1°C; the 2°C threshold would allow 0.6°C of additional warming. To put it mildly, neither of these targets will be easy to hit, given that temperatures have been rising by about 0.2°C a decade since 1990.

Avoid the Unmanageable, Manage the Unavoidable

Humanity has two options for containing temperature rise. The first option is to make sharp, continuing cuts in global emissions starting now. Pachauri, the IPCC chair, has said that global emissions must peak by 2015 and then fall rapidly until an 80 percent reduction, compared to 2005 levels, is reached by 2050. What's more, a sizable share of these reductions must come within the next ten years or the risk of runaway climate change increases. As previously noted, such dramatic reductions are technologically feasible and economically affordable. Investments in higher energy efficiency are particularly fast-acting and lucrative. We also need a crash program to reduce emissions of methane, a greenhouse gas that is twenty times more potent than carbon dioxide. Unlike CO_2, methane remains in the atmosphere for only a decade. Thus reducing methane emissions—from landfills, cattle farms, and other sources—has a much quicker cooling effect on global temperatures. If we want to stave off the melting of polar and glacial ice and other potential tipping points, argue former IPCC chair Robert Watson and United Nations Foundation senior fellow Mohamed El-Ashry, "methane is the most effective place for us to start."

A second option is to remove greenhouse gases that are already in the atmosphere or find ways to nullify their effects. After all, it is the total amount of greenhouse gases in the atmosphere, not the annual emissions, that determines how much global warming occurs. During the preindustrial era, the amount of CO_2 in the atmosphere averaged 280 ppm. After 250 years of industrialization, the level has risen to 390 ppm. More and more scientific studies are now concluding that the level must decline to 350 ppm or lower if we are to avoid catastrophic climate change. Pachauri (in his personal, not his IPCC, capacity), Hansen of NASA, and Nicholas Stern, former chief economist of the World Bank, are but three of the climate luminaries who have endorsed the 350 ppm target. But since the current level of 390 ppm has already surpassed it, and emissions are still rising by 3.5 percent a year, it is clear that the 350 ppm target cannot be achieved with emissions reductions alone: we must also extract CO_2 that is already in the atmosphere.

One way to do that is by utilizing photosynthesis and storing carbon in vegetation and soils. The earth's plants and soils are not yet removing enough CO_2 to halt rising temperatures, but they could do much more with proper stewardship. Currently available agricultural practices such as no-till farming could capture carbon dioxide equivalent to roughly 25 percent of annual global agricultural greenhouse gas emissions, according to the *Fourth Assessment Report*. Other methods for removing CO_2 or nullifying its effect—often referred to as geoengineering—include deploying mirrors in space to reflect some of the sun's rays away from Earth. But such schemes are highly controversial; some scientists warn that humans don't know enough about the earth's systems to undertake geoengineering safely—they might well make a bad situation worse.

"The geoengineering approaches considered so far appear to be afflicted with some combination of high costs, low leverage, and a high likelihood of serious side effects," John Holdren, President Obama's science adviser, said in April 2009. For example, some geoengineering proponents have suggested pumping large amounts of sulfur particles into the stratosphere, a higher layer of the atmosphere, just as volcanic eruptions do. Invisibly small, the particles released by the eruption of Mount Pinatubo in 1991 blocked enough sunlight to stall the global rise in temperatures for the following two years. But, critics point out, Pinatubo's eruption also appears to have substantially reduced rainfall, suggesting that "major adverse effects, including drought, could arise from geoengineering solutions."

White roofs are a better idea. Dubbed "geoengineering-lite" by Joseph Romm, a former U.S. assistant secretary of energy who blogs at climateprogress.org, the idea is to make roofs—and pavements—white, thus reflecting more sunlight away from the earth's surface at little or no risk. Citing research by Art Rosenfeld, the grand old man of California energy innovation, Romm reported that the average American household could counteract the ten tons of CO_2 it annually emits by retrofitting one thousand square feet of roof or sidewalk with reflective surfaces. Retrofitting all urban roofs and pavements in the world would yield emissions reductions equivalent to taking all the world's cars off the road for eighteen years.

These two options for reversing global warming—dramatic emissions cuts and sensible extraction of CO_2 from the atmosphere—are not mutually exclusive. Given the lateness of the hour, we will probably have to employ both of them. But it is worth emphasizing that there is no shortage of practical methods for cooling the planet; there are plenty more where these came from. What has been lacking is the will to put these methods into practice.

Our guiding strategy, I submit, should be the following: "Avoid the unmanageable and manage the unavoidable." I first heard this phrase from Madelene Helmer, a Dutch environmentalist who in 2001 convinced the International Committee for the Red Cross to make climate change a core part of its mission and who now runs the organization's climate center. But Schellnhuber says he originally coined the phrase, by chance, during a scientific meeting in Brussels in 2004 with a small group that included John Holdren of Harvard, who then spread it more widely. Schellnhuber said the idea could also be expressed in a more policy wonk way as "We must mitigate as much as possible and adapt as much as necessary." But I find the original formulation more accessible to average people. "To avoid the unmanageable," Helmer told me, "we must first recognize that there is such a thing as the unmanageable. If temperatures rise too far, climate change can make the earth, or at least large parts of it, unsuitable for human habitation. To keep from crossing that threshold, if we haven't crossed it already, we must cut emissions dramatically. At the same time, we must recognize there is such a thing as the unavoidable. No matter how much emissions fall, temperatures and climate impacts will increase for years to come. So we must cope with these impacts as best we can. And, I would add, we should pay special attention to helping the poor, since they will be hit hardest, despite having done nothing to cause the problem."

As the second era of global warming unfolds, it is clear that the old arguments that pitted adaptation against mitigation were a false choice. From now on, humanity must pursue both adaptation and mitigation at maximum speed. That means, on the one hand, that we must make our households, communities, companies, and countries *climate-friendly:* that is, they must be able to function while emitting few if any greenhouse gases. For if our emissions continue at anything like business-as-usual rates, temperatures and impacts will increase to where effective adaptation will become practically impossible. At the same time, we must make our countries and communities *climate-resilient:* as capable as possible of withstanding the impacts of climate change, which even in the best case will be substantial.

So how do we do that?

4 | Ask the Climate Question

Soon all sorts of strange things will come. No longer will things be as before.

—Folktale of the Wasco tribe of the Pacific Northwest

AS A FATHER in the second era of global warming, I have my good days and my bad days. The bad days you can probably imagine. Writing this book has taught me more than I'd like to know about our climate dilemma: about how drastically our civilization must change course to avoid catastrophe, how stubbornly some people and institutions resist even minor shifts in direction, and how destabilizing the impacts that are already locked in are likely to be. In the face of this, I make a conscious effort to avoid despair, for despair only warps thought and paralyzes action. Still, the analyses that come across my desk make it all too easy to envision some very dark outcomes.

But I have good days as well, and these are usually inspired by stories that show that the climate fight is not hopeless after all. One of my best days came in June of 2008, when I went to Seattle to interview Ron Sims. As the chief executive of King County, Sims was the top elected official of a municipality that encompasses the city of Seattle, some of its suburbs, and the corporate headquarters of Microsoft, Amazon, Starbucks, and Boeing. Over the past fifteen years, Sims had pioneered a fresh, farsighted, effective response to climate change that local governments across the United States and around the world were beginning to copy. He had linked his climate policy to a larger agenda of advancing social justice and pro-business economic development. And he had done this while remaining strikingly popular with voters, winning three straight elections by comfortable margins.

What most set Sims apart was the two-track climate strategy he employed. "We absolutely need to reduce greenhouse gas emissions, but we also have to adapt to the impacts we can no longer prevent," he told me outside his office in downtown Seattle. "The scientists say our region will see warmer, wetter winters in the future. The snowpack [atop the Cascades east of King County] will shrink. That means there won't be enough water for everyone if we don't get going on adaptation."

Although Sims's ecological commitment was ardent enough to earn him the nickname "Mr. Salmon," his argument for taking early action to prepare for climate change was based on tough-minded economics. "We think people and businesses will want to move to King County in the future because we took action to prepare for the world of 2050," he said. "We're taking steps to make sure we'll have enough water, we'll have levees that don't break, we'll have alternative energy sources, economic growth in the right places, a green work force. There are going to be winners and losers under climate change. I don't want King County to be a loser."

Many U.S. municipalities were setting themselves up for failure, Sims added, by continuing to support sprawling growth patterns. "The more sprawl there is, the more people drive and the more greenhouse gas emissions there are," he explained. "Sprawl is bad for adaptation, too. More sprawl requires a municipality to provide water and electricity across greater distances, which will be harder to do as water and energy become more scarce in the future." Recently, Sims had met the mayor of Atlanta, which, like much of the southeastern United States, had suffered record drought in 2007 and 2008. Mayor Mary Frances had said her city had "no zoning, no reclaimed water, and 900,000 housing units on septic," he recalled, his eyes widening in amazement. "That's a catastrophe in the making." Atlanta was hardly unique. "The counties around Dallas, Houston, and Phoenix that are allowing endless sprawl—those places are going to be in trouble in the future!" Sims exclaimed.

If sprawl is unsustainable, I replied, what should take its place?

"Our mantra is smart growth, green buildings, dense development, land preservation, and social justice," Sims said. "To cope with climate change, people must live more densely. But living more densely can create friction. So how do we do it safely? By attacking poverty and creating green, livable communities with less racial and class disparity. Greenbridge is a beautiful example of that."

Located a few minutes' drive from the Seattle airport, Greenbridge was one of King County's first green building projects. When Sims gave it the go-ahead in 2001, he told his advisers to "design the community we'll live in in the future." The job fell largely to Stephen Norman, a burly, wisecracking New Yorker whom Sims had recruited to run the county's Housing Authority. "When I got here from Manhattan, the first thing I had to do was learn how to drive," Norman told me. "It was a good lesson in how my built environment shaped my cultural expectations." Due to be completed in 2012, the $250 million Greenbridge project would consist of one thousand homes housing approximately 3,500 people, many of them recent immigrants.

Visually, Greenbridge was like no public housing development I had ever seen. Buildings were painted bright colors, came in different designs, and were bordered by attractive landscaping. Norman met me in front of the community center, where a roof of solar panels glinted in the sun. None of Greenbridge's residential structures—townhouses, cottages, and small-scale apartments—had solar panels yet, but Norman hoped they would before long. "We're building them 'solar ready,' so when solar panels get down to the right price, which shouldn't take too many more years, we'll be able to install them without tearing out walls," he said. Meanwhile, extra-efficient insulation would keep the buildings cooler in summer and warmer in winter, thus delivering both mitigation and adaptation at the same time, while also lowering energy bills.

In keeping with Sims's directive to "make it walkable," Greenbridge boasted parks, community gardens, and access to hiking and biking trails. But there was little grass. "No, you won't see much grass here,"

Norman said as we climbed concrete stairs to a block of rental housing where two little girls, one white and one black, merrily squirted one another with water pistols. "Grass is a terrible consumer of water, which gets expensive," he continued. "So we planted mainly drought-resistant trees and plants."

To protect against the opposite extreme of too *much* water, Greenbridge relies on *bioswales*—natural gutters made by digging an indentation into a flat strip of land. A bioswale runs a few feet to the side of a street; because it is about a foot lower, it collects storm water that might otherwise flood the street. Excess water flows down the bioswale to a collection area and is reused for irrigating parks and community gardens.

Soon, Norman and I had circled back to the community center, which contained a sparkling indoor basketball court and boys' and girls' club where kids could play or do homework after school. Next door was a library, an employment center, and an extension classroom for community college; out front was a spacious plaza. "This is the community gathering place," said Norman as we strolled across the plaza. "We'll have a regular farmers' market here so people can buy fresh fruits and vegetables. And underneath is a giant cistern, which will capture more storm runoff." Pointing across the street to a row of two-story buildings, Norman said the ground-level units would house commercial and retail shops, "so people won't have to drive a car to buy milk or have a coffee with a friend."

"Preparing a community against climate change isn't just about using green materials," Norman said. "It's about designing the built environment to integrate home, work, retail, and transit so people are less vulnerable to climate impacts and don't have to use cars as much." Norman cited a second housing project he had supervised, near the Microsoft campus in Redmond. "That's a very desirable part of King County, with lots of good restaurants and amenities, but who was going to wash the dishes and cut the grass?" he asked. "Entry-level workers couldn't afford to live within thirty miles of the place, so they were all driving their junkers every day, spewing pollution." The Village at Overlake Station was the first transit-oriented development of its kind in the United States, said Norman. Completed in 2002, it added 308 affordable housing units near the Overlake bus transit center. "We now average 0.6 cars

tional Association for the Advancement of Colored People. "When I was ten, an elementary school teacher told me, 'Your parents should behave themselves,'" Sims said. Smiling at the memory, he added, "When I got home from school that day and told my parents what happened, my father looked at my mother and said, 'I guess it's time for another demonstration.' Two weeks later, we had the demonstration, and I never heard another word from that teacher."

At age sixty, Ron Sims had flecks of white in his closely cropped hair but the unwrinkled face of a much younger man. He was compact, with a thick neck and shoulders, and he spoke and moved quickly, exuding a restless energy. Staffers wondered when he slept; it was common to receive e-mails from him at three and four o'clock in the morning. "I get an idea in my head and I can't sleep until I get rid of it," Sims explained with a grin.

One of Sims's ideas was to make climate change central to the mission of every department in county government. "Ron is always telling us, 'Ask the climate question,'" said Jim Lopez, Sims's deputy chief of staff. "That means, check the science, determine what conditions we'll face in 2050, then work backwards to figure out what we need to do now to prepare for those conditions."

Fortunately, scientists at the local University of Washington had been asking the climate question since 1995, when Professor Edward Miles founded the Climate Impacts Group (CIG). "Climate change is sure to occur in some form," declared the CIG's first report, *Impacts of Climate Change on the Pacific Northwest*, issued in 1999. "At present, many natural-resource-dependent commercial enterprises and government agencies operate on the assumption that climate is unchanging. It will take considerable effort to replace that assumption. . . ." Subsequent CIG studies sketched potential impacts on the Pacific Northwest—on temperatures, water, soil, forests, pests, diseases, and plant and animal species—for the 2020s, the 2040s, and 2080s. What astonished Miles was that Sims actually read the reports. Not only that, he began basing policy on them. "Can you imagine that?" Miles asked. "An elected official who cares about the year 2050, not just the next election, and who takes appropriate action? He stands out from the herd."

Sims was not the only local public official who had done extraordinary things about climate change. Greg Nickels, the mayor of Seattle, actually had a higher national profile, thanks to his role in rallying hundreds of cities around the United States to reduce greenhouse gas emissions. In February 2005, when the Kyoto Protocol became law in the 141 countries that ratified it—a list that did not include the United States—Nickels announced the U.S. Mayors' Climate Protection Agreement. Under the agreement, cities would voluntarily pledge to meet or beat the protocol's targets of reducing emissions approximately 7 percent below 1990 levels by 2012; the mayors also promised to urge state and federal governments to do likewise. The idea caught fire: by 2010, more than one thousand U.S. mayors representing 86.6 million citizens had joined the effort. Then Nickels went further, pledging to reduce emissions in Seattle by 80 percent by 2050—matching the IPCC's recommendation for how to avoid the worst scenarios of climate change.

For his part, Sims had begun pursuing emissions reductions from the time he first became county executive in 1996. He made King County the first county government in the United States to join the Chicago Climate Exchange, a voluntary trading system that uses market mechanisms to foster emissions reductions; the county was also the first to use hybrid engines in public buses. By 2007, the King County government had reduced the emissions from its own operations by 6 percent below 2000 levels. The city of Seattle had done significantly better, reducing emissions of the entire city (i.e., the public, private, and household sectors) by 8 percent below 1990 levels.

Sims realized that government alone could not win the climate fight; society as a whole—the business community, schools and churches, the general public—had to get involved. So in October 2005, he and Ed Miles convened a conference in Seattle called "The Future Ain't What It Used to Be: Planning for Climate Disruption." It was the first major conference in the United States to focus on preparing for the impacts of climate change, and one of the first in the world. Sims made sure his message was hard-hitting without being demoralizing. After listing the various threats posed by climate change, he shifted gears and called the climate crisis "the greatest opportunity our society and world has ever faced. If we do what it takes to reduce greenhouse gas emissions to safe levels and prepare for the impacts we see are under way, we will transform the economic foundation of modern civilization and . . . realize better health, social justice and sustainable economic development throughout the world."

By all accounts, "The Future Ain't What It Used to Be" was a great success. "We were planning the conference long before Gore's movie [*An Inconvenient Truth*] came out, so we thought there might be 100 people there," Sims recalled. "But we had 650 show up. People were ready to hear what they could do." Afterward, Sims and his staff compiled a guidebook so others could learn how to create and manage their own adaptation programs. Written in collaboration with Miles and his colleagues at the CIG, the 186-page manual, *Preparing for Climate Change*, describes in clear, nontechnical prose why to start adapting now, how to prepare and implement a climate action plan, what pitfalls to avoid ("Expect surprises" reads one item), and other valuable information. Available at http://www.icleiusa.org, the guidebook is useful for households, businesses, and nongovernmental institutions as well. For anyone wishing to tackle the nuts and bolts of adaptation, *Preparing for Climate Change* is an excellent place to begin.

"A Levee Breach Here Would Cost $46 Million a Day"

The fairy tales told by the native peoples of the Pacific Northwest often focus on the natural world: how the mountains, rivers, sky, and land came into being; which animals and spirits enjoyed special powers; how the native peoples fit into the larger scheme of things. In *Indian Legends of the Pacific Northwest*, scholar Ella E. Clark collected more than one hundred tales that for generations had been transmitted orally—told to groups gathered around fires, accompanied by chants, mimicry, and other dramatic effects on the part of the storyteller. Clark found "striking parallels" between these legends and the early literature of Europe, which was likewise "filled with the deeds of giants, monsters and superhuman heroes."

Perhaps the most striking parallel is to the Bible; many of the Pacific Northwest's tribes told stories bearing an uncanny resemblance to the Old Testament account of the Great Flood that Noah survived by building an ark. As in the Bible, the native stories included one or more morally upstanding people who, joined by children (but not animals), were loaded into a gigantic canoe to ride out the storm while bad people were left to perish. That so many such stories exist suggests that flooding was a recurrent aspect of ancient life in the Pacific Northwest. Then as now, storms blew inland from the ocean and traveled east until they collided with the Cascade Range, where they dumped precipitation in the form of rain or snow. The Cascades mark the great climatic divide in the Pacific Northwest: they separate the wet west from the arid east. The massive mountain visible from modern-day Seattle has long been known to white people as Mount Rainier, but native peoples called it various versions of *Tkomma,* a word that lives on today as the name of the nearby city of Tacoma. All the native peoples, noted Clark, "referred directly or indirectly to the streams of white water coming down [the mountain's] slopes."

Modern science tells us that these streams of white water will grow larger and more volatile as climate change intensifies in the years to come. According to the CIG's studies, mean temperatures in the Pacific Northwest will increase by 1.9°F by the 2020s, by 2.9°F by the 2040s, and by 5.6°F by the 2080s. These higher temperatures will cause more precipitation to fall as rain rather than snow and also cause more of the snowpack to melt. As a result, more water will flow down the mountains in winter and spring, increasing the likelihood of flooding in the flatlands that stretch westward through King County to the sea.

One day, I went with Mark Isaacson, the director of King County's Water and Land Resources Division, to see some of the improvements Sims had ordered made to prepare against flooding. As Isaacson nosed his car into midmorning traffic on Interstate 5, he said we were heading south of Seattle to the Green River, whose snow-fed flow provides most of the water for southern King County. A few minutes after passing the airport, we pulled off the interstate and descended a long hill. Soon we were rolling through a zone of low commercial buildings separated by half-empty parking lots. It was hard to believe this was some of the most economically valuable real estate in King County. But a levee breach here, Isaacson said, would cost the local economy $46 million a day.

"Many of these buildings are warehouses that supply food and other critical goods to Seattle," he explained. "Two semi trucks and ten to twenty small trucks a day deliver from here to grocery stores. Restaurants receive 1,200 deliveries a day. Starbucks has a big distribution facility here, quite a few medical supply companies, too. If a levee broke, the roads here would be underwater, and all those deliveries would stop." The levees protected 65,000 jobs that generated $3.7 billion of income a year, Isaacson added.

Farmers had built levees along the Green River fifty to sixty years ago, said Isaacson, but those levees were little more than mounds of earth extending along the riverbanks. They were sufficient to protect farmland that could afford to flood occasionally, but inadequate when billions of dollars of commerce and modern infrastructure were at risk. Like all the departments in King County government, Isaacson's had been told by Sims to ask the climate question. Once they did, Isaacson said, "My colleagues and I knew right away that we had to upgrade our levees. The problem is, that gets really expensive. Our budget was nowhere near big enough. The only way I could see it happening was with a tax increase, but I was very reluctant to suggest that. I'm a public servant, I take my responsibility to the taxpayers seriously, and I know politicians would rather have a root canal than tell voters, 'I'm gonna raise your taxes.'"

But when Isaacson outlined the problem, Sims didn't flinch. "Ron told me, 'We have to do it. But we have to explain to people *why* their taxes have to go up, why it's in their interest that these improvements get made.' And that's pretty much what happened. My staff outlined a program of levee improvements and calculated that the cost would average $40 per household in the Green River valley region. Then we reached out to mayors of towns in the valley and to the public. We had open meetings where we explained the situation. People didn't grumble much. Even towns not located right next to the river agreed to pay because they understood that their economic well-being would suffer if the levees broke."

The tax increase, approved by the voters in 2007, increased Isaacson's budget tenfold. Instead of the $3.4 million per year he had received in the past, the flood control program was allocated $335 million over the next ten years—monies to be used for repairing some five hundred levees and revetments in the county's flood defense system.

Pulling into a parking lot, Isaacson invited me to see one of the first repairs financed by the taxes. We parked facing a grassy, gently sloping hill about eight feet tall, which turned out to be the backside of the Briscoe School Levee. A railed set of stairs took us to the top of the levee. Below us was the Green River, about thirty yards wide, flowing from left to right. My visit took place in June, so the water level was low, about twenty feet from the top. "In winter peak flows, it's within a foot of the top," said Isaacson. "We'll have twelve thousand cubic feet per second of water coming through here then."

The top of the levee was paved with asphalt for the benefit of hikers and bikers. Stepping to the edge, Isaacson pointed to a ramp of dirt that led halfway down to the river before flattening into a shelf that extended about twenty feet into the river. "We put that shelf of dirt and rocks there to deter erosion and stabilize the levee," he said. "Even when the shelf is underwater, it will slow the river's flow. We also added those logs along the banks. They'll deflect the water away from the levee toward the middle of the river and provide good habitat for salmon. And we set the levee back—in effect, we widened the river—so more water can flow through here."

"How many other parts of the system need similar repairs?" I asked.

Pulling a thick stack of papers out of a manila folder, Isaacson said, "Here's a list of current projects. We prioritize them according to relative risk—how damaged a given levee is today and how great the consequences of failure would be. We've made a good start, but we've got a whole lot of work to do in the next ten years."

"I Can't Put My Head in the Sand"

If the ancient tales of the Pacific Northwest are any guide, preparing against floods may be the easy part. Native peoples appear to have passed down not a single story concerning drought. But then the Pacific Northwest is famous for its frequent rains, at least west of the Cascades. The future will be different.

There will be much less water available as climate change intensifies, according to the CIG, which warned in a 2004 report, "Adaptability to drought presents one of the greatest challenges for Pacific Northwest water resource systems." The CIG projected that *overall* precipitation will not change dramatically, increasing 2 percent a year by the 2020s, another 2 percent by the 2040s, and a further 6 percent by the 2080s. But higher temperatures will cause more of the precipitation to fall as rain rather than snow while also melting the snowpack. Thus river flows will be greater in winter (hence the likelihood of winter flooding) but lower in summer. On the positive side, the increased flows will allow the region's many hydropower plants to produce more electricity in winter. But a corresponding decrease in summer flows will mean less hydropower in summer—just when electricity demand will be rising as the extra heat boosts reliance on air conditioning.

Delivering sufficient water for households, agriculture, and industry under these conditions will be difficult, the CIG declared, and humans are by no means the only ones at risk. A melting snowpack also threatens forests, the scientists pointed out. Less snow will mean drier soil, making it harder for seedlings to sprout. A projected increase in wildfires and pests—again, because of higher temperatures—could kill off still more trees. A shrinking snowpack would also be bad news for many animals, especially salmon, a fish of enormous cultural and economic importance in the region. Smaller river flows and higher river temperatures would make it harder for salmon to reproduce, the CIG said, warning that "the prospects for many Pacific Northwest salmon stocks look bleak."

As Sims saw it, the task of government under such circumstances was to prepare people and institutions to live with less water. That meant, first of all, convincing them that the problem was real. "People didn't want to believe there were going to be water shortages," he recalled. "After all, this is a place where it always rains. But I said, 'This is what the science says. We have to respect it.' The reason we have so many ecological problems today is because we didn't listen to science."

In the American West, the traditional response to water shortages has been to go out and find—or steal—more of it. But the shrinkage of the snowpack makes that unlikely. In theory, reservoirs could be built to capture the snowmelt before it flows downstream and disappears into the Pacific. But the CIG was pessimistic: most of the region's river basins already contain all the reservoirs they can accommodate. The CIG cautioned further that implementing *any* strategy will be difficult because the region's water management system lacks a single authority with the power to execute decisions. (Management of the Columbia River alone rests with eight federal agencies, thirteen tribes of native peoples, multiple state agencies in seven states, numerous public utility districts, and hundreds of government subdivisions in the United States and Canada.)

Taking the scientists' recommendations to heart, Sims proposed a set of initiatives that respected ecological realities but upset bureaucratic tradition and popular sensibilities. Rather than seeking to increase the *gross* supply of water, he fought to maximize the *net* supply. He did so both by using forestland as a natural reservoir and, most controversially, by reusing wastewater before it was released to the sea. The latter idea provoked a fierce political battle that eventually had to be settled by the state legislature.

The morning we met, Sims took me to the site of one of the toughest fights in that battle, the Brightwater wastewater facility. The idea behind recycled water is simple: instead of using pure water for all human purposes, why not substitute recycled water for watering golf courses, irrigating landscapes, and supplying factories? The Brightwater facility would take in wastewater, run it through filters to remove contaminants, then pump it out for delivery to nonhousehold customers. In effect, using reclaimed water would allow the county to use the same volume of water twice, doubling the amount of water at its disposal. That sounded unobjectionable until one mentioned the so-called yuck factor: the reclaimed water had previously been used to wash people's dishes, fill their bathtubs, and flush their toilets. Sims said most people, however, got past this problem: "We explained that reclaimed water would be carefully filtered and never used for drinking, bathing, or irrigating crops." The real objections, he continued, as we drove east from Seattle, were economic. "The golf courses don't mind reclaimed water," he said. "The pushback came from water agencies that had been selling the golf courses water. One of the [agency] people asked me, 'Do you know how much money we make from golf courses?' It's money! We have to get past the question of who's making money on things and do what's right for the community as a whole."

Water agencies resisted Sims, accusing him of a power grab aimed at stealing their business. The former mayor of Seattle, Paul Schell, also complained, charging that Sims's proposal would raise the price and lower the quality of water in the city. The state legislature eventually joined the fray, blocking the plan from proceeding for three years while it debated who owned the water in question and what state policy on reclaimed water should be. But in the end, Sims triumphed. What was his secret?

"I just didn't back down," he replied. "King County had the right of approval over the water planning process and we exercised that right. And we had the science on our side, which was crucial. I told water agencies, 'We're going to have less water in the future. You may not like it, but that's a fact.' As an elected official, if I know what's coming, I can't put my head in the sand and wish it weren't true. I have to listen and act."

The Brightwater plant is located twenty miles from Seattle on 114 acres of wooded land in neighboring Snohomish County. The site used to house an auto junkyard, but the wrecks had been cleared away as part of a program to protect nearby Little Bear Creek and add forty acres of hiking trails. "We'll leave this place in much better shape environmentally than we found it," said Sims.

After bouncing up a rocky driveway, our vehicle stopped above a huge construction site. Below us, bulldozers snorted exhaust while workers in hardhats hoisted rebar. "This plant relies on an advanced membrane bioreactor treatment technology, the most ecologically friendly filtration system in the world," said Gunnar Goerlitz, the project manager. "When it opens in 2011, it will be the largest plant of this type in the world, capable of treating 36 million gallons of sewage a day." Of the 36 million gallons treated, 15 million will receive only secondary-level treatment and be pumped through a tunnel into Puget Sound. The remaining 21 million gallons will receive additional treatment from the bioreactors, which separate solids and bacteria from water molecules, and be distributed to end users. Some of the water will go to a golf course near Microsoft's campus in Redmond, but most will be added to the Sammamish River, boosting the flow of agricultural water in the area.

Gunnar confirmed that the Brightwater plant had encountered strong opposition: "When the planning of this facility began in 1999, lots of people didn't like the idea of including purple pipe [the color used for reclaimed water]. When we got to the design stage in 2002, we asked whether we should put in purple pipe. Ron said, 'You bet.' It cost $26 million, but it would have cost much, much more if we had to come back later and retrofit it, because you'd have to dig up all the tunnels again."

Looking down on the construction site, Sims smiled with undisguised satisfaction. "When people in the year 2050 look at this plant, they'll say, 'Those old-timers did this right,'" he said. "Every other water treatment plant is going to have to be retrofit. Not this one. We did this one right."

The Ripple That Creates a Tsunami

Ron Sims was not a preacher's son for nothing; spreading the gospel of adaptation was a moral issue to him. In 2006, the San Diego Foundation invited him to participate in the city's "Understanding Global Climate Change" lecture series. "Near the start of my speech," Sims recalled with a mischievous grin, "I told them San Diego was a beautiful place but King County was going to beat San Diego in the race for the twenty-first century. They looked kind of shocked. I explained that in King County we were planning *now* for the conditions we'd face from climate change and other global trends by 2050. In the question-and-answer session, someone got up and asked how San Diego could do that. I said they first needed to get a handle on the science, so why not send some of their scientists up to speak with the folks at the University of Washington about doing an impacts study? The person wanted to know how much that would cost. I said probably a couple hundred thousand dollars. He said, 'Well, I'm in for fifty thousand.' Then someone else said, 'I'm in, too.' They ended up sending a team from the Scripps Institution [of Oceanography] to talk with us. They did their own study and, eventually, their own adaptation program."

"Ron's speech was the watershed event that shifted the mood in San Diego about what climate change meant and how it could affect our community," said Linda Pratt, who played a key role in the city's adaptation efforts as manager of the government's Sustainable Community Program. "It was our *Inconvenient Truth* moment, you might say. Ron got a standing ovation afterwards; people were asking him to run for president."

Sims had one question he wished every local official would ask: "'In an age of global warming, what is your community going to look like?'

If my peers would ask that question, they would see the need for adaptation. Don't look at the national or the global level; look at your community level. Ask your local scientists what the impacts will be where you live. And once you have the answer, the question is, 'If you're a public official and won't do anything about important information you have, why are you in office?'"

More local officials have begun embracing adaptation in recent years, and for an obvious reason. Local governments are the "first responders" to droughts, heat waves, and other impacts of climate change, explained *Ask the Climate Question,* a report the nonpartisan Center for Clean Air Policy in Washington published in 2009. The report described how ten North American municipalities, led by King County and including Toronto, New York, Chicago, and San Francisco, were beginning to pursue adaptation. Many of the measures were first steps: recognizing the need to do adaptation in the first place, incorporating it into existing procedures, commissioning scientists to project impacts. But the point was to get the process started, said Steve Winkelman, who coauthored *Ask the Climate Question* with Josh Foster and Ashley Lowe—to break through the previous indifference or prejudice against adaptation. "I think we're past the point now of people seeing adaptation as surrender," he said. "What's becoming clear is that you need to do both, adaptation and mitigation. It's like eating and breathing: you can't do just one."

Is King County's record on climate change perfect? Hardly. Levees are being upgraded, but only to a 1-in-100-year flood level. Drought remains a major threat. Notwithstanding the Brightwater wastewater facility and Sims's other innovations, long-term trends of water supply and demand in the Pacific Northwest are out of sync, and climate change will make things worse. Going forward, either supply must expand or demand decline or both. In any case, it's hard to imagine that many salmon will survive. Heat waves are another problem. "There is a perception that heat waves won't affect us because temperatures won't reach 120°F like they will in Phoenix," said Elizabeth Willmott, Sims's global warming coordinator. "But we don't have the capacity to deal with 90 to 95°F days on a regular basis. We have cooling centers for short-term relief, but in the long term we need green buildings that keep people cool without air conditioning." King County may also be underestimating how much sea levels will rise. Relying on CIG projections, Sims told the Port of

Seattle and other stakeholders that they should raise piers and other infrastructure enough to accommodate thirty-seven inches (one meter) of sea level rise by 2100. Actual sea level rise, as noted earlier, may well be higher.

Nevertheless, Sims pronounced himself satisfied with efforts to date. "The climate science of 2020 will know more than we know today," he told me. "We tell the Port, 'At least prepare for [thirty-seven inches of] increase, and if science later says it's more, at least we'll be partway to the goal.' The next generation will have to be more sophisticated, but at least we've gotten things started."

"I'm very hopeful," Sims told me as we bade farewell. "I think we're about to see an incredible acceleration in adaptation around the world. We've heard from public officials in Chicago, London, Stockholm, France, Spain, even Thailand, inquiring about our work. We're glad King County can be an example. We want to be the ripple that creates a tsunami of adaptation."

Finding the Sweet Spots

Nowhere else has Sims's message had greater influence than in Chicago, that most American of cities, with its beguiling combination of urban grit and midwestern friendliness, commercial muscle and cultural sophistication, sparkling wealth and chronic poverty, world-class universities and ethnic neighborhoods, boastful competitiveness and farm belt work ethic. As a brawny commercial giant and the hub of America's transportation network, Chicago could hardly be more different from the leafy, liberal Pacific Northwest of King County. Yet those differences are precisely what make Chicago's "very, very aggressive" climate plan potentially more influential than King County's, argued Sadhu Johnston, who served as Chicago mayor Richard M. Daley's point person on climate change. "If Cleveland or Philadelphia hears that Seattle or San Francisco is gearing up against climate change, they aren't sure they can follow," said Johnston, Chicago's chief environmental officer from 2003 to 2009. "But here's Chicago, with our manufacturing heritage, industrial infrastructure, and brownfields, our enormous size, and we're taking on climate change. If we can do it, and show that it makes your city better, that's a model for the world and people will listen."

Johnston and his colleagues made no secret of their debt to Sims and King County. Speaking of the adaptation guidebook Sims and his staff compiled, Joyce Coffee, the executive director of Chicago's climate change office, exclaimed, "That book was my bible, and I know the same is true for my counterparts in Milwaukee and other midwestern cities." Johnston said that King County was "constantly a model for us—on their climate plan, on green procurement, storm water management, and a lot of other issues. We always look at what they're doing. We don't copy it, but we do try to build off of it. We all need places to learn from."

Not that Sims would have minded, but Johnston came close to outright copying when he introduced an early draft of Chicago's Climate Action Plan in November 2007. Speaking to a roomful of community and environmental activists, Johnston began by calling climate change "an unprecedented problem but also an unprecedented opportunity"—exactly what Sims had told his 2005 conference—before adding, "We believe that addressing climate change can help us to increase the number of jobs in Chicago and improve the quality of life." Also like Sims, Johnston gave adaptation equal billing with mitigation, emphasizing that the "Building a Greener Chicago" plan would include "steps both to reduce emissions and to prepare for changes that can't be avoided." The mitigation goals included 25 percent cuts from current levels by 2020 and 80 percent cuts by 2050; adaptation would focus on preparing for intensely hot weather and more severe storms.

Chicago's climate plan went far beyond what any other major American city had proposed; indeed, it compared favorably with practices in Britain. "It was very impressive how they put adaptation right up front," said Chris West of UKCIP, who attended the briefing because he was in town for the Chicago Humanities Festival. "I would only expect that from cities in Britain we have directly worked with, such as London, Manchester, and Glasgow. Any city or company has to have a champion for adaptation, and Chicago clearly has one, who can stand up at a board meeting and say, 'This is something important to the core mission of this community or corporation.' Otherwise, it's not going to be taken seriously."

The adaptation champion in Chicago, according to Johnston, was the mayor himself. The environment had been a big part of Daley's governing image since he launched a series of high-profile green initiatives in the 1990s, including installing a green roof on City Hall. Johnston, whose boyish face made him look a decade younger than his thirty-four years, gave me a tour of the roof at the start of our interview. As we emerged from a dark stairwell into the open air, he named various grasses and shrubs that dotted the roof's surface, including a pretty ground cover called sedum that was known as "the camel of plants" because of its low water requirements. Surrounding us were the skyscrapers of Chicago, a city that had "the most green buildings in the world," according to Johnston, who pointed out one after another in the skyline. "Some mayors see [green initiatives] as a matter of saving the planet," Johnston continued. "The mayor sees that as a benefit, but his fundamental goal is to improve the quality of life in Chicago. That's what makes his message so powerful, in my opinion."

In 2003, after installing the green roof at City Hall, cleaning up the once-putrid Chicago River, initiating a citywide recycling program, and greening the procurement policies of city agencies, it occurred to Daley to ask the climate question.

"The mayor uses Blue Notes, which are short memos on blue paper, to communicate with his staff," said Johnston. "A Blue Note can be as specific as 'There's a tree on the corner of streets X and Y that needs a trim,' to big cosmic inquiries. About a month after I started this job, the mayor sent me a Blue Note asking, 'How is climate change going to affect Chicago?' I had to respond that I didn't really know. We went back and forth on it a while and finally decided we had to pull in some serious science." To co-chair the effort, Johnston invited Adele Simmons, one of Chicago's civic leaders, who had been following climate change since heading the MacArthur Foundation in the 1980s, where she funded some of the first climate projects of U.S. environmental groups. Together, Johnston and Simmons organized a team of experts, many from the University of Illinois, to project the impacts of climate change on Chicago and to analyze mitigation options and their economic implications. The project began in December 2006; Mayor Daley officially unveiled the city's climate plan in April 2008. (Details are available at http://www.chicagoclimatechange.org.)

"When we got the impacts study back, it was actually a big relief," Johnston said. "We're better off in Chicago than most major cities will be. It turns out there are advantages to not being on a coast. We expect no serious water problems. Our two major concerns over the next fifty years are going to be frequent extreme summer heat and more severe storms, especially in winter."

But more extreme heat is no small matter: Chicago is already very hot and muggy in summer. The city suffered the deadliest heat disaster in modern U.S. history in 1995, when "a blend of extreme weather, political mismanagement, and abandonment of vulnerable city residents resulted in the loss of water, widespread power outages, thousands of hospitalizations and 739 deaths in a devastating week," Eric Klinenberg wrote in an article adapted from his book *Heat Wave: A Social Autopsy of Disaster in Chicago.*

City officials were determined to prevent a repeat of the 1995 disaster. By 2100, said Johnston, Chicago could experience thirty days a year when the temperature exceeds 100°F (40°C), compared to three days a year currently; on seventy days a year the temperature will exceed 90°F (35°C), compared to twelve to fifteen currently. The increased heat will endanger more lives, especially among the poor, sick, and elderly, and place extra demands on the city's first responders—the ambulance, fire, and police officers who tend to distressed residents. Johnston said the city had improved its emergency response plans after the 1995 disaster, but "now we'll upgrade those plans." The extra heat will also stress electricity supplies and worsen air quality. "Chicago already has the highest [per capita] asthma rate in America," said Coffee. "What will we do when it's even hotter? We can tell people not to go outdoors, but above all we've got to lower ozone pollution levels."

Air conditioning—one obvious countermeasure—raises two problems from a climate perspective. In the short term, increased use makes outdoor temperatures even hotter (air conditioners cool indoor spaces by expelling heat outdoors). Second, unless it is powered by wind, solar, or other noncarbon energy, more air conditioning means more greenhouse gas emissions and additional global warming.

Air conditioning is a classic example of a key policy challenge in the second era of global warming: finding adaptation techniques that do not make mitigation harder. A second example of an adaptation policy that runs counter to mitigation is seawater desalinization, which is often mentioned as a response to deeper droughts. But desalinizing seawater requires enormous amounts of energy. If that energy is produced from fossil fuels, it will make global warming worse and thereby contribute to even deeper droughts in the future. Instead, what is needed are adaptation techniques that have a positive—or, at worst, a neutral—effect on the mitigation imperative.

Johnston called this challenge "finding the sweet spots" between adaptation and mitigation. Chicago was pursuing two such sweet spots in response to fiercer heat waves: planting trees and shifting to wind-powered electricity.

"We've started to map this stuff," Johnston told me, unfolding a series of detailed city maps on a worktable in his City Hall office. The first map showed so-called urban heat islands—parts of the city where temperatures were markedly higher than elsewhere. A second map charted density of tree cover. When Johnston overlaid the second map on the first, the areas of low tree density often overlapped with the areas of high temperatures. But from now on, said Johnston, the city would target its tree planting at these heat islands, seeking to drive down temperatures. "In the past, we planted trees in an ad hoc manner," he told me. "An alderman [a member of the Chicago City Council] or a citizen would call up and say, 'Could you plant some trees on our street?' And we'd come out and do it. Now we're going to target the urban heat islands, which often"—now he overlaid a third map on the first two—"are areas populated by lower-income people, who tend to be more at risk from heat waves." In addition, said Johnston, the city planned to plant trees in fifty acres of scattered empty lots along the highway that connects downtown and O'Hare International Airport, lots that otherwise were wasted space.

Chicago also aimed to become America's capital of wind power manufacturing. Already, eight of the world's leading manufacturers had chosen Chicago as their North American headquarters, said Howard Lerner, the director of the Environmental Law and Policy Center in Chicago and a key contributor to Chicago's climate plan. After rattling off the companies' names and nationalities—including E.On of Germany, Suzlon of India, and Iberdrola of Spain—Lerner explained, "They're here because the Midwest is the Saudi Arabia of wind power. There are 25,000 megawatts of wind power now under development in Illinois, Minnesota, Iowa, and the Dakotas." Chicago offered the transportation infrastructure, manufacturing facilities, skilled labor, and positive policy environment these companies needed to sell wind turbines and related equipment to customers throughout the United States, Lerner said. "The blades of modern wind turbines are two hundred feet long. They're not like refrigerators you can manufacture in China, put on a container ship to California, and truck over the Rocky Mountains to sell at Best Buy in Peoria. That's why wind manufacturing—not just turbines but gear boxes, switches, ball bearings—is coming to the Midwest."

Chicago hoped to use wind power to meet both economic and climate protection goals. "In 2007, Illinois enacted, thanks in part to our center's work, the strongest renewable energy standard in the United States," said Lerner. "By 2025, 25 percent of the state's energy must come from renewables, compared to 2 percent in 2008. In its climate plan, the city endorsed that goal as well. Now, it's up to city agencies to make it happen."

Mayor Daley also harnessed Chicagoans' habitual competitiveness to push his green agenda, said Johnston. "The way we started the Green Hotels Initiative was, the mayor invited the general managers of thirty leading hotels to breakfast. He said he wanted Chicago to have the greenest hotels in the world, and green hotels would win special recognition from the city. Every one of those hotels signed on, and twenty-five of them won Green Seal certification. We've found that if you make it a competition, it just takes off. So now we're doing that with Green Museums, Green Campuses, Green Stadiums, you name it."

"Making it fun is one of the ways we'll get the public involved in the climate plan," added Suzanne Malec-McKenna, the director of the city's Department of the Environment. "The message can't be gloom and doom. That turns people off. But if you tell them how they can help and maybe even make money in the process, they're eager to get started."

Daley also drew on his political connections to address perhaps the main obstacle to improved energy efficiency: a lack of up-front investment capital. Former president Bill Clinton came to Chicago in November 2007 to announce with Daley that the Clinton Climate Initiative philanthropy was committing $10 million to finance efficiency improvements in Chicago. Buildings are responsible for roughly 40 percent of America's greenhouse gas emissions, so improving their efficiency promises real progress on mitigation. Since better efficiency also saves money through lower fuel bills, efficiency investments seem a no-brainer. The problem is, the money savings accrue over time, while the costs of insulating roofs, upgrading old appliances, and replacing leaky windows have to be paid up front. That was a deal breaker for many building owners, who did not have the necessary funds. The Clinton Climate Initiative offered a way around the impasse. The initiative had gotten private banks to contribute a reported $5 billion to finance up-front costs of improving energy efficiency in buildings; Chicago was the first of twenty-two cities around the world where the funds would flow.

"The way it works is, an ESCO [energy services company] audits a building and says, 'Do these ten things and it will save, say, $100,000 a year in energy costs,'" Johnston said. "But doing those ten things will cost $450,000. So you pitch the idea to a bank and the bank says, 'We'll loan you $300,000.' Then the city would loan the additional $150,000 needed to make the project happen. The energy savings [$100,000 a year] are used to repay the loans over the next five years." In the first year of the project, Johnston added, Chicago would retrofit three hundred units at a cost of $5 million to the banks and $2.5 million to the city; if that went smoothly, the program would be dramatically ramped up in future years.

Meanwhile, the city was experimenting with more advanced methods of green practices. Johnston and I left City Hall and walked a few blocks to an alleyway that led from the Goodman Theatre to the Chicago Theatre. At first glance, the alley didn't look unusual. But Johnston drew my attention to ground level. The sidewalk looked like normal cement, but in fact it was a substance known as permeable concrete that would help cope with the intense storms climate change would bring: rain would seep into this substance rather than flash off into storm drains and increase flooding. "We're also experimenting with a special photocatalytic cement that was developed for the Vatican to keep the Millennium church white," Johnston said. "Apparently this cement has a chemical reaction that causes it to eat smog. We're trying out these ideas at our first 'green street,' in the Pilsen neighborhood on the southwest side of Chicago, to see what works. Infrastructure—the streets, lights, drainage systems—makes up 25 percent of Chicago. That is a huge green opportunity."

Interviewing Johnston the day before the 2008 presidential election, I asked if he would be moving to Washington to pursue a green agenda as part of the Obama administration. "No, cities are where it's at," he replied. "Cities are the economic engine of the country. They're where 80 percent of the population lives, so city governments can have a huge impact on environmental issues." Then a smile creased his boyish face and he added, "But it will be really nice to have some federal assistance for a change."

Little did Johnston know that some of the federal assistance would come from none other than Ron Sims. Although Sims had backed Hillary Clinton in the Democratic presidential primaries, President Obama chose him in early 2009 to be the deputy secretary of the Department of Housing and Urban Development. In accepting the number-two post at the $39 billion agency, Sims said he looked forward to helping America "prepare for the age of global warming," adding, "Success can only come if we transform our major metropolitan areas." He soon showed what he had in mind. In June 2009, HUD joined with the Department of Transportation and the Environmental Protection Agency to announce a Partnership for Sustainable Communities, an initiative that closely parallels policies Sims had championed in King County. From now on, the announcement said, the federal government would "provide more transportation choices"—read, "encourage mass transit more than private vehicles." It would also "promote equitable, affordable housing"—read, "encourage developments like Greenbridge"—and "support existing communities"—read, "stop subsidizing sprawl." True, announcing a policy in Washington is just the beginning of making it happen on the ground across the country. But the Partnership for Sustainable Communities represents a 180-degree shift in direction for the federal government. If Sims and his new colleagues can make good on their promises, they may inspire the tsunami of adaptation he yearned for after all.

There's No Silver Bullet, Only Silver Buckshot

Richard Daley was not the only U.S. big-city mayor following in Sims's footsteps. Mayor Michael Bloomberg of New York City actually was slightly ahead of Daley in launching a comprehensive climate action plan. In a speech he delivered on Earth Day 2007 to hundreds of government, business, and community leaders at the American Museum of Natural History, Bloomberg declared that coping with climate change was imperative to New York's future. As a coastal city, he pointed out, New York was particularly threatened by "rising sea levels and intensifying storms." Climate change would also worsen New York's already ferocious summertime heat and humidity and stress its water and energy supplies. Bloomberg urged facing these problems "not in the future, not when it's too late, but right now." Toward that end, his speech outlined a long-term sustainability plan for the city, a plan he called PlaNYC.

Bloomberg's remarks focused overwhelmingly on mitigation—he pledged to cut the city's carbon footprint by 30 percent by 2030—but PlaNYC included adaptation as well. Having become a billionaire before his election as mayor in 2002, Bloomberg clearly grasped the economic appeal of energy efficiency, and he made it the centerpiece of his mitigation strategy. Most of the projected cuts in greenhouse gas emissions would come from increases in energy efficiency, including retrofitting buildings and mandating the purchase of more efficient lights and appliances—what the mayor called "Spend an extra dollar today, save two tomorrow." More reductions would come from encouraging cleaner electric power generation. He also advocated expanding subway and bus service and discouraging private cars by imposing a congestion fee on driving in midtown Manhattan. On the adaptation side, Bloomberg said the city would plant a million trees over the next three years. Not only would the shade of these trees reduce street temperatures, thus making them a form of adaptation; they would also reduce air pollution and absorb carbon, thereby aiding mitigation as well.

One of the most expensive proposals in PlaNYC was the investment it proposed to make in the aging system of aqueducts and tunnels that bring New York its water. The vast majority of New York's water originates in the Catskill and Delaware river watersheds, hundreds of miles away. Two massive tunnels deliver this water to the city. But neither tunnel had been inspected for more than fifty years; authorities literally had no idea what shape they were in. What they did know was that a failure in either tunnel would leave millions of New Yorkers without water. They also knew that climate change would increase the stress on the water system, because the northeast of the United States was projected to experience more volatile rainfall in the years to come, and this would produce larger pulses of water pouring through the tunnels. PlaNYC's solution was to urge the completion of the long-planned but always-postponed Water Tunnel Number 3. Completing the tunnel would cost billions, but it would enable temporary closure of Water Tunnels 1 and 2 so engineers could inspect and modernize them. This was an investment the city could not afford not to make, Bloomberg argued.

Completing Water Tunnel Number 3 illustrates how adaptation is a win-win proposition, said Cynthia Rosenzweig, a senior research scientist at the Goddard Institute for Space Studies at Columbia University and the chief science adviser on New York's plan. "Adaptation helps you manage today's climate extremes—the storms and floods that would be occurring regardless of climate change—as well as the greater extremes that climate change will bring in the future," Rosenzweig told me.

Despite having spent more than twenty years investigating climate change, Rosenzweig was relentlessly cheerful, with graying blond hair and a chipper voice. After beginning her career as an early member of James Hansen's research team at Goddard, she now ranked as one of the world's leading experts on adaptation to climate change. "For twenty years there has been a tremendous focus within the IPCC, as there should have been, on the science of climate change and methods of mitigation," she said, speaking in her office at Goddard of NASA, one floor below Hansen's office. "Now, we're in a new phase, a phase of not just searching for solutions but testing them. And there will be many, many solutions. There's no silver bullet, only silver buckshot."

Under Rosenzweig's leadership, New York City had established what she called "a local version of the IPCC" to provide ongoing scientific advice to decision makers in both the public and the private sectors as they refined and implemented the vision of PlaNYC. Like the IPCC, the New York City Panel on Climate Change was composed of experts from a wide range of physical and social sciences; it issued its first report in February 2009.

"We'll offer advice on the levels of protection needed for infrastructure through 2080," Rosenzweig told me. "Let's say the MTA [Metropolitan Transportation Authority] is renovating the Wall Street subway station, which is in a flood zone. Today, they wouldn't be thinking of adaptation to climate change. But going forward, our panel's reports will tell them the level of flooding they have to protect against." Or, to be more precise, what *levels* of flooding they have to protect against. Reflecting the uncertainty that plagues most analyses of potential impacts of climate change, Rosenzweig and her colleagues decided to devise two separate projections for future sea level rise: a standard projection and a so-called Rapid Ice Melt Scenario projection. Their standard projection concludes that New York must prepare for seven to twelve inches of sea level rise by 2050 and twelve to twenty-three inches by 2080—a considerable but manageable amount. But if, as Hansen and some other leading climate scientists feared, the earth's polar ice melts at a rapid rate in the years ahead, the Rapid Ice Melt Scenario will tell New York infrastructure managers that they could experience nineteen to twenty-nine inches of sea level rise by 2050 and forty-one to fifty-five inches by 2080—a much more challenging scenario.

"We won't tell the MTA or the airports how to [achieve that level of protection]; that's up to their staffs," Rosenzweig continued. "Maybe they'll raise the runways. Maybe in the early years [of sea level rise] they'll decide to just close the airport for a day or two during high storm surges. But they will know these issues must be dealt with."

Officials cannot make prudent decisions without also improving their knowledge about New York's existing risks, said Adam Freed, the city government's deputy director of Long-Term Planning and Sustainability. "We need much better flood maps," Freed told me in April 2010. "Our current maps have a margin of error of three feet. So when Cynthia's group projects two feet of sea level rise by 2080, or by 2050 under the Rapid Ice Melt Scenario, our current maps could be massively underestimating the total land area that could flood. Or they could be massively overestimating it. We just don't know." To rectify the problem, the city has begun deploying an aerial technology known as LIDAR (Light Detection and Ranging). Airplanes are flown over the city's landmass and a special laser pulse is sent to earth that measures a given area's elevation much more precisely than the technologies that inform flood maps compiled by the United States government's Federal Emergency Management Agency (FEMA).

Like King County, New York City sought to involve a cross section of the community in its climate action. PlaNYC was informed by more than fifty public meetings where citizens and business and educational leaders were invited to give suggestions. To learn how best to engage and work with city residents on adaptation solutions, PlaNYC launched a pilot project in Sunset Park, a low-income, ethnically mixed neighborhood on the waterfront in southwest Brooklyn. Lessons learned there would in turn be translated into guidance for other neighborhoods. The city partnered in Sunset Park with UPROSE, a Latino community group that focuses on environmental justice. The city did so, said UPROSE's executive director, Elizabeth Yeampiere, because officials "recognized that they can't do adaptation without people getting involved at the grassroots level. You can do all the Al Gore slide shows you want, you won't move the issue forward unless local people feel a sense of ownership."

When Michael Oppenheimer mentioned parts of New York that three feet of sea level rise could submerge, Sunset Park was among the first he pointed to. UPROSE had proposed an ingenious yet practical response to the problem. For years, Yeampiere told me when I visited her office one morning, UPROSE had wanted to create a twenty-five-acre park along a section of the Brooklyn waterfront that was currently inaccessible, thanks to a phalanx of old warehouses and unused naval facilities that stood in the way. She drove me to the site in question, weaving her car through a dense warren of back alleys, past signs that indicated the area was off-limits. We emerged onto a concrete pier that extended about one hundred meters into New York Harbor and boasted a killer view of the harbor, the Statue of Liberty, and the skyscrapers of lower Manhattan. "A waterfront park here would give a real boost to the quality of life in this neighborhood," Yeampiere told me, "and it would make sense from a climate adaptation perspective, too. This land is too prone to flooding to be used for commercial or residential buildings. But it doesn't matter if a park floods once in a while."

Yeampiere saluted city officials for electing to meet with Sunset Park residents, who, she said, gave them an earful. "There's always a danger when experts come into the room that community people will defer too much because they have less education," she told me. "So we had our own meeting in advance to talk through the issues. People just blew me away with their creativity and insights. These are people who would be our scientists, our engineers, our government planners, if circumstances in their lives had been different. We compiled a list of things that we could do and a list that the city could do, too. We realized that a big storm could put our local hospital underwater, so we reached out to the hospital's management about preparing against such impacts. We're also asking whether we could use our churches as sanctuary if there were a disaster and our government failed to respond, as happened after Hurricane Katrina."

Nevertheless, mitigation remained the Bloomberg administration's top climate priority, and its officials made no apologies for that. "I'll be honest: I fully support adaptation, but I want it to have a lower profile in order to keep the pressure on for aggressive mitigation," said Rohit Aggarwala, the director of the city's Office of Long-Term Planning and Sustainability and the mayor's point person on PlaNYC. "I've gone to a lot of public meetings about this plan and I can't tell you how many people have said, 'Why don't we just build a seawall around the city?' They see a seawall as a silver bullet that's preferable to cutting back on driving and other mitigation steps."

Aggarwala, a clean-cut New Yorker in his mid-thirties, knew that a seawall might well be necessary someday, but he also knew that it was far from an ideal solution. For one thing, it would be hugely expensive—easily tens of billions of dollars—and take decades to complete. Even then, it would inevitably leave parts of the city unprotected. Oppenheimer, for example, had speculated that one segment of seawall could be built across the Throgs Neck channel, connecting northern Queens to the mainland. A second segment could connect western Brooklyn to Staten Island, and a third segment could link Staten Island to the New Jersey shoreline. But even this ambitious design would leave much of Staten Island and Brooklyn outside the ring of protection.

And what about all the other impacts of climate change? A seawall would be no help against harsher heat waves. Thomas Frieden, then New York City's public health commissioner (and later head of the national Centers for Disease Control and Prevention under President Obama), worried that stronger heat waves could cause coliform contamination of seawater and force the closure of city beaches. "And closing beaches is not a good idea when you have hundreds of thousands of people who want to go to the beach," he said.

But mitigation has proven a difficult struggle for New York City. One reason is the state government, which in 2008 voted down Bloomberg's proposal for congestion pricing after the New York City Council had approved it. That defeat in turn undermined Bloomberg's proposed expansion of mass transit, which was to be financed in part by congestion fees. The state government had also declined the city's request for a sixfold increase in energy efficiency funding.

"Changing the carbon footprint of a city as big as New York is like changing the direction of a supertanker," Aggarwala told me, and individual New Yorkers weren't making the job any easier. "Plasma TVs take three times more power than conventional TVs, and more and more people are buying them," Aggarwala added. "And air conditioners! New Yorkers used to have an air conditioner in one room of their apartment. Now, it's in two or three rooms. So we have a million more [air conditioning] units in New York than ten years ago. That's a steep curve to climb."

Aggarwala seemed more optimistic the last time I interviewed him, in April 2010. New York's electricity consumption per capita had finally fallen in 2008, the most recent data that was available. True, 2008 had been the start of the global financial crisis and ensuing economic recession, and recessions invariably reduce energy use. But Aggarwala cited studies the city had done that showed consumer behavior was changing as well. He also said that energy audits of commercial buildings in New York undertaken as part of PlaNYC had led some owners to order retrofits that had improved energy efficiency.

On the adaptation side of the challenge, Aggarwala was also encouraged that "the level of understanding of the likely impacts of climate change among the people who manage the city's critical infrastructure, both public and private, has increased tremendously, and that alone has reduced our risk. Those people are making much better decisions now." For example, a new power plant being built in Sunset Park was being elevated four feet above its original design level to cope with sea level rise. The Parks Department was planting only trees that could manage the heat and precipitation conditions anticipated in the future.

Aggarwala did confess to one lingering frustration, though. "We have to get people to understand the difference between prevention and resilience," he said. "Some people think we can keep climate change out somehow; you just build a seawall or a dome or something. You can't do it. Subways, for example, you can never perfectly protect. They are below sea level by definition and you can't seal them because you need the heat down there to get out. Instead, our goal has to be to increase our resilience, to get our people and infrastructure through whatever impacts occur as smoothly as possible. But lots of people just want a quick and easy fix. It's the same mentality as buying a Hummer that runs on bio-diesel—the idea that if I just change this one thing, I can fix the problem. It's delusional. And insanely frustrating."

How Individuals Can Make a Difference

Because my daughter was such a powerful motivation for me to write this book, I got into the habit of asking people I interviewed if they had kids and, if so, how they—as parents—dealt with the implications of climate change. Joyce Coffee, the city of Chicago's global warming coordinator, said she often saw the future as dark, so of course she worried about what that meant for her three-year-old son. I mentioned how Chiara's obsession with *The Nutcracker* had unwittingly opened my eyes to the hopeful messages in fairy tales. "I never thought of that," she said, "but it makes a lot of sense. My son Andrew's hero now is Spider-Man, and I've seen how he uses Spider-Man in difficult situations. When he faces challenges, like not wanting to go to school or dealing with me on limits he doesn't like, he'll take on Spider-Man's persona. He tells me, 'Momma, even if it looks scary, you have to step off the building and shoot your web shooter at the bad guy. Then it'll be okay.'"

Fortunately, one need not actually step off buildings to build a brighter climate future. The average U.S. household could reduce its greenhouse gas emissions by 25 percent within six months by making a few changes in daily routines, such as walking or biking to work instead of taking the car, according to the King County government. Bear in mind that the IPCC has estimated that greenhouse emissions must fall by 25 percent by 2020 if humanity is to have a fair chance of avoiding catastrophic climate change. If every household in America followed King County's advice, the United States could reach that target in six months rather than ten years. What's more, most of the recommended changes are good for you: they either fatten your wallet or trim your waistline, or both. (Details are available at http://www.kingcounty.gov/exec/globalwarming.aspx.)

As for adaptation, an individual person, family, community, or company can begin the process just like King County (and New York and Chicago) did: by researching and identifying one's vulnerabilities to climate change impacts. Start by asking yourself where your water supply comes from. This can actually be a fun family or school project: starting with your kitchen and bathroom faucets, trace the flow of water into your house back to its source. Maybe the source is a backyard well, but more likely it is a water treatment plant or reservoir and, before that, a river. Is that river fed primarily by snowmelt or a spring? How secure is that source likely to be in the future? That raises the question of whether climate change is expected in your geographical area to lead to an increase in rainfall, a decline, or both. Of course, you need to investigate your vulnerability to other impacts as well. How likely are future heat waves? If you live near a coastline, how far above sea level is your house or apartment building? Will the region's roads be passable in the event of a foot or two of sea level rise combined with storm surges? Remember: a sixth-floor apartment may not be much good if the trucks that deliver food, medicines, and other necessities can't reach your local shopping area. Don't forget peak oil either. It is highly likely that gasoline and other petroleum-based fuels will be harder and costlier to obtain in years to come, so the less dependent you can be on cars, highways, and other technologies that require petroleum, the better off you'll be. Some of the data needed to answer these questions may not be readily available, but don't give up: governments and research institutions are publishing more and more of it all the time.

The next step is to put in place proper safeguards against your vulnerabilities. If heat is a problem, think about planting shade trees near your home and painting the roof white. If the water supply is iffy, learn how to live with less. That could involve installing more efficient toilets and showers, adjusting personal habits, and changing landscaping to favor drought-resistant plant varieties, as was done in the Greenbridge housing development. If, like one friend of mine in the San Francisco Bay Area, you own a house on the water, in a spot that is not technically or financially feasible to protect with a seawall, your wisest course of action may be—sorry—to sell the house soon, while there are still people willing to buy it.

Still, as valuable as individual actions are, they change only so much.

When I asked Sims about being a parent under climate change, he said he had three grown sons, aged thirty, twenty-eight, and twenty-two. He figured they would be able to look after themselves, but not *by* themselves. "If society as a whole does not act, even valiant individual efforts will have only limited effect," Sims explained. Only government, he continued, can make sure that adequate sea defenses are built, water supplies are protected, and health systems are kept up to date. In the end, there is no substitute for government action.

Governments, however, usually must be pushed to act by citizens. Individuals must join together to demand action from their public officials, and they must make it clear that a failure to respond will be punished at the ballot box. Otherwise, most elected officials simply will not act. So, if you live in a community that is *not* emulating what King County, Chicago, and New York are doing to deal with climate change—and at this point, that includes most communities in the United States and overseas—you and your neighbors can ask your public officials why not. Urge your local media to ask them, too. You can start a discussion in your community about why reducing emissions and preparing for impacts is only common sense. Ask local scientists for help. Download copies of King County's adaptation guidebook and share it with local officials, business leaders, and heads of local schools, churches, hospitals, and other community centers.

Whatever form your personal and political involvement takes, the final imperative is this: don't wait to get started. As we'll see in the next chapter, the Dutch are doing more than anyone else in the world to safeguard against the impending impacts of climate change. Their cardinal rule? Begin now: you have farther to go than you think.

5 | The Two-Hundred-Year Plan

There is absolutely no reason to panic, but we must be concerned for the future. If we are to be well prepared for the expected consequences of climate change, we shall have to strengthen our flood defenses and change the way our country is managed. . . .

—Sustainable Coastal Development Commission,
the Netherlands, 2008

SINCE BEGINNING WORK on this book, I have often thought about where Chiara should live when she grows up and I am perhaps no longer around to look after her. Every community on earth will be affected as climate change intensifies in the years ahead, and its impacts will only be growing stronger by the 2020s, when Chiara will be about to make her own way in the world. Clearly some places will be safer than others. Should Chiara stay in the San Francisco Bay Area, where she was born and raised, despite the substantial impacts climate change will have here? Or should she perhaps move someplace else?

After seeing what Ron Sims and his colleagues have done, I put King County at the top of my mental list of possible relocation spots for Chiara. For much the same reason, I would also consider Chicago or New York. To be sure, none of these places will have an ideal climate twenty years from now. The Pacific Northwest, Sims told me, will be even cloudier and rainier in the future than it is now; the summer heat in both Chicago and New York will be ferocious. But challenging weather extremes will be the rule, not the exception, for many places in the future. The American Southwest in particular I would stay away from. With scientists projecting sustained record droughts for the region, water shortages seem all but certain, especially considering that most local governments are doing little to get ahead of the problem. By contrast, King County, Chicago, and New York are actively preparing themselves for a hotter, more volatile climate. If "Avoid the unmanageable, manage the unavoidable" is the new imperative in the second era of global warming, these three localities have made a good start on what remains, inevitably, an unfinished agenda. Each is aware of the new realities of climate change; each is asking the climate question; each is seriously working the problem. For all the mistakes they will surely make along the way, the fact that they have started early can only be a good thing.

I'm also keeping my eyes on a place even farther away: the Netherlands. In fact, I have applied for Chiara to receive dual U.S. and Italian citizenship, partly so she will have the right to study, work, and live in the countries of the European Union, including the Netherlands. I know: at first glance, the Netherlands does not seem like the safest place to ride out the next fifty years of climate change. After all, much of the country is already below sea level. Well below sea level. When visitors land at Amsterdam's Schiphol airport, their aircraft touch down on runways that are 14.5 feet lower than the nearby North Sea. That is much lower than the country roads that access Chiara's California beach town, lower even than the salt-laden rice fields around Uma's village in southern Bangladesh.

But the Dutch, I believe, have the most impressive plan in the world for adapting to climate change. The plan is well thought out, well funded, and supported by world-class technical knowledge and infrastructure. Above all, it benefits from its social context: the Dutch have a long history of coping with floods, storms, and other forms of water stress, and they do so with an extraordinary degree of collective cooperation that is nevertheless utterly unsentimental. True, climate change adds a substantial new wrinkle to the challenge, but the Dutch are not panicking. They are even making sure that foreign investors and tourists understand that their country is not surrendering to climate change. "When we decided in 2006 to make a national plan for adapting to climate change, we created a slogan to signal the rest of the world that we plan to be here for a long time to come," said Pier Vellinga, the nation's leading climate scientist. "It says, 'We Are Here to Stay.'"

In climate change as in real estate, it is location, location, location that matters. Like the American cities of New Orleans, New York, Washington, and Boston, as well as such overseas counterparts as Tokyo, Buenos Aires, and Alexandria, the Netherlands owes its extreme vulnerability to climate change to its delta location at the confluence of a great river, or rivers, and the sea. The Dutch government itself has called the country "the drain of Europe" because it is the place where some of the continent's largest rivers, including the Rhine, at last spill into the ocean. Rising temperatures are already shrinking the Alpine glaciers that feed the Rhine; as temperatures rise further, the shrinking will intensify, causing more and stronger downstream flooding. The North Sea poses even greater dangers, because North Sea storms arrive with much more force—and much less warning—than river floods do. It takes five days for a flood to work its way down the Rhine from Germany to the Netherlands, whereas a North Sea storm can rise to a ferocious strength within hours. Climate change will make the North Sea more dangerous by raising sea levels and possibly strengthening storms. If a storm manages to breach the Netherlands' fabled sea defenses, the country might not survive. "Seventy percent of the Dutch GNP is earned below sea level, which is also the place where most of our people live," the Dutch crown prince, Willem Alexander, has said. Since the Netherlands is one of the most densely populated countries on earth, evacuation would be, to put it mildly, extremely difficult.

The Dutch are savvy enough to realize that outsiders might consider their nation too precarious to warrant protection. "If you view the problem of climate change adaptation from the perspective of Europe as a whole, you might say, 'Let's forget that low-lying area in the northwest of the continent. It's no use trying to save the Netherlands,'" said Aalt Leusink, a senior adviser to the government. "But we look at it from the

Dutch point of view. This is our country. We want to stay here. And we have decided, in the face of climate change, that we will try to stay here for the next two hundred years."

That's right: the Dutch are planning two hundred years ahead as they adapt to climate change (and some of their scenarios even gaze four hundred years ahead). Relying on scientific studies by Vellinga and others, the government commissioned Leusink in 2006 to coordinate the adaptation plan, and already significant strides have been taken. The Dutch are spending large amounts of money, making tough decisions, and engaging both the public and the private sector at every step of the way.

Outlandish as it may appear to outsiders, taking a two-hundred-year perspective on climate change is entirely in keeping with Dutch history. After all, the Dutch have been defending themselves against flooding rivers and stormy seas for more than eight hundred years already; archaeological evidence shows that a dam had been constructed across the Amstel River, the site of modern Amsterdam, by 1275. The Dutch have even manipulated water levels to defeat foreign armies. "In the eighteenth century, we used our water system to pester the French and Spanish armies out of our country," Vellinga told me. "When they came with horses and wagons to try to take our land, we deliberately flooded those areas—but only enough so the water was two feet deep. That's not deep enough to go over in a ship, but it is too much for a vehicle to pass through, so they would get stuck. It worked very well."

Climate change promises to be at least as challenging an enemy, said Vellinga, who has ranked among the world's leading climate scientists since he helped establish the IPCC in 1988. "The world is in for at least 2 degrees [3.6°F] of temperature rise if we get our act together, or 4 degrees [7.2°F] if we don't," Vellinga said. "And we'll *feel* it with 2 degrees. Even 2 degrees are enough to cause about one meter [three feet] of sea level rise, which is no small thing."

"It Is Essential to Start Early"

"You can either adapt to climate change because you are forced to or because you plan it," Leusink told me. "We propose to plan it." He had come to fetch me at the train station in The Hague so I could see for myself some of the adaptation initiatives he was overseeing. Trim, with a pale complexion beneath whitening hair, Leusink had spent most of his life working in the private sector, running and advising companies; in the 1990s he served on the board of directors for Schiphol airport. Now he was a consultant for both public- and private-sector clients. But "I am independent, not a political appointee," he emphasized. "I am free to say what I think."

As we climbed into his midnight-blue Audi, Leusink told me that preparing his country against climate change would be a massive task. "Climate change will touch every issue and every place in the Netherlands," he said. "Obviously our sea and river defenses will be affected, but so will the rest of our water management system, our transportation, agriculture, land use planning, the construction business—everything." Harsher heat waves, for example, will stress the public health system. Leusink and his colleagues wanted the system strengthened, and they were also advising more creative steps. They wanted buildings to be "climate-proofed" to handle extreme temperatures better; proposals included using more insulation and painting external surfaces light colors that will reflect rather than absorb the sun's rays. They also wanted more trees planted to offset the "urban heat island effect"—the tendency of cities, with their heat-absorbing concrete and metal surfaces, to get hotter than nearby rural areas.

"For adaptation to be successful, it is essential to start early," Leusink continued as we exited the station and headed east under hazy summer skies. "That is especially the case in the Netherlands because we have to take a *lot* of measures and for every measure we need physical space. For example, we want healthy ecosystems, so species must have space to move from one place to another. But it takes time to free this space from its current uses."

In barely ten minutes, we had reached the countryside. Coming from the United States, I couldn't help but notice that we had gone from dense downtown to open farmland without passing through the unsightly sprawl of gas stations, fast-food joints, and shopping malls that surrounds American cities—sprawl that, as Ron Sims emphasized, runs counter to both adaptation and mitigation. "We can't waste space here," explained Leusink. "The Netherlands is a very crowded country. Physical space is the most important constraint on our ability to adapt to climate change. To find more space, we have decided to reduce the amount of land devoted to agriculture. So we will produce less cheese and milk in the future. That's okay. These products are easy to import from Denmark and Germany."

"But isn't cheese important to your economy?" I asked. "Gouda cheese is one of the signature brands of the Netherlands, isn't it?"

He chuckled. "Of course it is important to protect important brands like Gouda. But we can maintain that brand without producing the actual milk needed to make the cheese. I'll tell you a secret. Ninety percent of the herring sold in stores as Dutch herring is actually imported from Denmark and Norway."

Leusink turned the car onto a one-lane road that led past flat green fields dotted with sheep and Holstein cows. On one side of the road, a shirtless man was raking long strands of golden hay in the midday heat. On the horizon, about two miles away, stood three ancient windmills, all in a line. Nicknamed "The Three Men," the windmills had provided the power that was used to pump this area dry, turning what had been a shallow lake into usable farmland. The land was then protected against future flooding by surrounding it with mounds of packed earth. The Dutch called the mounds *dikes* and the drained farmland a *polder*. Such landscape engineering had been a common practice in the Netherlands for centuries. Beginning in 1533, it was how the Dutch had settled—one can almost say created—the western third of the country, which previously had been too swampy to inhabit.

Now, Leusink said, this long-standing practice would be reversed in order to adapt to climate change.

"This is a polder we will flood," Leusink said, pointing at the vast field that lay before us. "Two years from now, it will become a lake again so it can provide water storage. Scientists tell us that under climate change our rainfall will become more erratic. We expect the Netherlands to receive somewhat more rain per year than now, but the key point is that this rain will tend to fall in short, heavy bursts rather than in gentle showers. And there will be long periods when there is no rain. So we have decided to create lakes. These lakes will absorb the downpours so we don't have as much flooding and also so we have a supply of fresh water in times of drought. We also plan to upgrade our sewage and storm drain systems to deal with more torrential rains. This will cost about 15 billion Euros [$23 billion]."

"But what about the farmers who live here?" I asked. "What happens to them if this land becomes a lake?"

"We will buy them out," said Leusink. "They will be paid a fair price."

"What if they don't want to be bought out?"

"They have no choice," he replied. "This decision was made in a coordinated fashion by the provincial government, the federal government, and the national water board. We had an open, democratic process. We spent four years talking it through at the local level. But finally you must act. You cannot allow one or two people to block an action that is best for everyone else."

"Take Account of Adaptation in All Aspects of Life"

The most famous Dutch fairy tale is a tale the Dutch themselves don't much care for: the story of a little boy who stuck his finger in a dike, thus saving his town from washing away in a flood. The Dutch are quick to point out that this story was written by an outsider, the American writer Mary Elizabeth Mapes Dodge, who clearly knew nothing about how dikes work. "It's the silliest thing in the world to think that putting a finger in the wall of a dike would hold back surging water," said Dano Roelvink, a professor of coastal engineering at the UNESCO-IHE Institute for Water Education. "The pressure of the water would instantly blow the hole open. This story is known in Dutch schools, but it is known as legend, not fact."

Yet a similar act of heroism, on a much larger scale, saved many lives during the greatest natural disaster in modern Dutch history: the floods of 1953. On February 1 of that year, a very heavy North Sea storm combined with an abnormally high spring tide to send massive waves crashing over and through the nation's dikes. "All of southwestern Holland was flooded," Roelvink said. "Rotterdam was barely saved when a small-town mayor ordered a ship captain to steer his boat into a hole in the dike." But this act of bravery only kept a bad situation from getting worse. The 1953 floods devastated the Netherlands, killing an estimated

1,835 people and leaving 72,000 homeless, while smashing sixty-seven large breaches and some four hundred holes in the nation's dikes. The influx of salt water ruined 200 hectares of agricultural land for the next five years.

Much as Hurricane Katrina would be in 2005, the 1953 flooding was a nationally traumatizing event that sparked widespread soul-searching and urgent calls for reform. The difference is, the Dutch followed through with genuine, comprehensive reforms, notwithstanding the associated difficulties and expense. "The 1953 floods were *the* event that sparked a transformation in the nation's consciousness and approach to coastal protection," said Roelvink, who explained that a committee of "wise men" was appointed to formulate a plan "to protect the Netherlands from the threat of the sea forever." The so-called Delta Committee proposed a radical overhaul of the flood protection system. The national government would take over responsibility from local authorities. Safety standards would be raised exponentially. Henceforth, the dikes and dams protecting against North Sea storms had to be able to resist a 1-in-10,000-year storm surge—in other words, a storm so great that it had only a 1 in 10,000 chance of occurring each year. River dikes had to be able to resist a 1-in-1,250-year flood. These extremely high standards dwarfed the 1-in-100-year level that prevails in most countries of the world even today. The Dutch needed such a high standard, the wise men argued, because in a country so low-lying, a failure to contain flooding could easily yield catastrophic consequences.

To achieve the higher safety levels, the Dutch embarked on one of the great engineering feats of the twentieth century: construction of the Delta Works, a series of gargantuan dams and barriers in southwestern Holland that closed off most of the rivers and channels through which storm surges could reach inland. (Across the English Channel, where the 1953 floods were also very destructive, the authorities responded by, among other measures, approving construction of the Thames Barrier east of London.) Because rivers still had to flush to the sea and carry cargo traffic, some Delta Works structures were designed with movable parts; the Maeslant Barrier, which protects Rotterdam, swings open and shut like a butterfly's wings. It took more than thirty years to complete most of the Delta Works, and the Maeslant Barrier was not activated until 1996, forty-three years after the terrible floods that called it forth.

The Dutch thought they had solved the problem, but in the mid-1990s they found out differently. "In 1993 and in 1995 we got hit with two big floods that led to hundreds of thousands of people getting evacuated," recalled Henk van Schaik, a water engineer with the Co-operative Programme on Water and Climate of UNESCO, based in the city of Delft, the epicenter of Dutch water management. "The public said, 'Bloody hell, government—do something!'"

One of the people who answered the call was Pier Vellinga. Lean and self-confident, with graying hair swept back from a broad forehead, Vellinga had been a driving force behind Dutch climate policy since the late 1980s, when he helped organize the scientific meetings that led to the creation of the IPCC. In the 1990s, he began working for the Ministry for the Environment, where he soon found himself tangling with the nation's top meteorologist, who was not persuaded that climate change was much of a threat. The 1993 and 1995 floods, along with the ongoing work of the IPCC, shifted the argument in Vellinga's favor. Nevertheless, almost ten more years passed before the Netherlands fully embraced the challenge of adapting to climate change, with Hurricane Katrina supplying the final, decisive push.

"The renewed awareness of the risks facing our country started at the intellectual level in 2003 with the Erasmus Lecture I gave on the safety of the Netherlands in the view of climate change," said Vellinga. In the lecture, Vellinga discussed the latest research he and one of his PhD students, Laurens Bouwer, had just completed. "We found that the impacts of climate change would make the risk of flooding much bigger than Parliament had agreed to," Vellinga recalled in our interview in the old canal town of Utrecht. "Our old assumptions were based on historic patterns that no longer held. National policy previously assumed we should plan for 60 centimeters [2 feet] of sea level rise by 2100, but now I believe that needs to increase to 1.5 meters [5 feet]."

But these scientific concerns did not gain political traction in the Netherlands until Hurricane Katrina struck two years later. "The basic issue after Katrina was 'Could this happen in the Netherlands?'" Vellinga continued. The answer, as supplied by a team of Dutch scientists chaired by Vellinga, was "We are safer than New Orleans but also more vulnerable, [because] much more capital and many more people are at stake [here]. Moreover, in 2005 a significant part of [our country's] flood defenses did not meet the legal safety standard."

In 2006, the Dutch government formally embarked on a national adaptation plan. Responding to urgings from Vellinga, one of the first steps taken was the establishment of a second Delta Commission. Like the commission of "wise men" that charted a new course for the country after the 1953 floods, the new Delta Commission (officially, the Sustainable Coastal Development Commission) would combine scientific and engineering expertise with political and economic concerns to forge a fresh, common vision for securing the country's future. There was one major difference, however. "We have been asked to come up with recommendations not because a disaster has occurred, but rather to avoid it," the commission wrote, adding that its goal was "to present an integrated vision for the Netherlands for centuries to come."

Centuries to come indeed. In a world where politicians rarely think more than four years ahead and banks and corporations focus on quarterly returns at best, the Dutch have deliberately crafted a convincing two-hundred-year plan for adapting to climate change. Relying on research conducted by Vellinga, Michael Oppenheimer, and many others, the second Delta Commission declared that the Netherlands has to prepare for 0.65 to 1.3 meters of sea level rise by 2100 and 2 to 4 meters of sea level rise by 2200. This in turn requires that "the level of flood protection must be raised by at least a factor of ten with respect to the present level," a level that, as noted, is already the highest in the world. Securing the nation's freshwater supply was also expected to be a major challenge and requires taking "direct actions . . . now."

Equally remarkable, the Dutch private sector has actively supported the adaptation program and even agreed to pay some of the costs. Although corporations in many countries, especially the United States, have often pressured government to weaken actions against climate change, the business community in the Netherlands accepts that the threat is real and the need for action clear.

According to Leusink, Dutch business leaders intervened in the debate in part because they wanted the government to clarify what steps would be taken so that businesses could plan accordingly. "In the past, government planning documents usually forecast twenty-five years ahead and did not take climate change into account," he told me. "But businesses wanted a view of government policy beyond twenty-five years, because the investments they were making will last longer than that."

As a businessman himself, Leusink argued that adaptation must henceforth be elevated to the heart of all public and private decision making. "It should become as integral as a financial analysis," he said. Whether it is a policy shift by government, an investment by a private company, or a purchase by an individual citizen, a decision must be considered in light of how it affects and is affected by the necessity of adapting to climate change. "This is our core message," said Leusink, his tone growing unaccustomedly intense. "We have to take account of adaptation in *all* aspects of life. Don't make a special adaptation policy for your business, or create a new ministry for adaptation within government. You must mainstream adaptation so that it becomes part of *all* policies and actions."

"In Places, We Will Retreat"

Adapting to climate change will take plenty of money, but the Dutch seem prepared to spend it. "I did extensive research at the government's request and found that by spending 0.2 percent of our GDP, about 1.3 billion Euros a year [$1.95 billion], we could maintain our current level of risk in the face of one meter of sea level rise," Vellinga told me. To cope with the high-end climate change scenarios (including 2 to 4 meters of sea level rise by 2200) will require doubling the annual investment to about 0.4 percent of GDP, or 4 billion Euros ($6 billion) per year. On a per-capita basis, that means each Dutch person will pay some 240 Euros ($360) a year for flood protection—about the same amount they pay now, Vellinga added, for fire insurance.

The core of the Dutch adaptation plan focuses on flood protection, with different approaches for the threats posed by rivers and by the sea. New protection standards will be established by 2013. The engineering, construction, and administrative work needed to achieve them in practice are to be completed by 2050.

To cope with larger river flows, the Netherlands is implementing a "Space for the River" policy that, like the polder flooding Leusink showed me, will require some households and infrastructure to move. Instead of seeking safety by trying to control the flow of water, "Space for the River" accepts that achieving such control will be increasingly difficult as climate change intensifies; therefore, the more prudent course is to find a way to live with increased water flows. In practice, this requires giving up some land so water has more room. "Historically, the highest flow of the Rhine has been twelve thousand cubic meters of water per second," Vellinga said. "At the moment, our dikes are built to handle fifteen thousand cubic meters per second, but for the future we have decided the level must be eighteen thousand cubic meters per second. So we will widen the banks to enable the river to absorb 20 percent more water."

One area slated for river widening is near Nijmegen, a city in the extreme east of the country, close to the German border, that straddles the Rhine. As if to illustrate Vellinga's earlier history lesson, the city ranks as the oldest in the Netherlands. Apparently, Nijmegen was as far north as the Romans got; they settled here in 6 A.D. Now, parts of the outskirts of the city will become what locals I interviewed disparaged as "sacrifice zones." In the event of excessive flows down the Rhine, the areas of Nijmegen called Ooijpolder and Duitsland will be allowed to flood in order to protect lives and property in the more densely settled areas to the west. The inhabitants of these two sacrifice zones protested bitterly to the government but were overruled.

"Some people in Nijmegen did complain," Vellinga confirmed, "but I testified before Parliament and said they would be crazy to listen to these complaints. The people in Nijmegen argued that if there is a big flood on the Rhine, it will flood first at Cologne [about eighty miles upriver, in Germany] and the water coming to the Netherlands would then be lower. 'Why should we flood,' they asked, 'when Germany will get hit first anyway?' My argument to Parliament was 'Do you think Germany will get hit twice? We need some *realpolitik* here. Once they get hit by a bad flood, they will build dikes and then we will be hit even harder.' This episode, for me, demonstrates that as we adapt to climate change, we must estimate not only the behavior of rivers and coasts but also the behavior of neighboring countries."

The greatest danger comes from the North Sea, Vellinga stressed, and here the Dutch plan of improving flood protection by "at least a factor of ten" will proceed in phases. In phase one, he said, "we are making sure to take care of the weakest spots in our coastal defenses right away." Much of the Dutch coast, from Rotterdam northward past Amsterdam to Den Helder, is protected by sand dunes. These dunes are the country's first line of defense against ocean storms, and until I saw them with my own eyes I didn't understand how they could fulfill that function; where I live in California, dunes tend to be relatively short and broken by gaps that would allow waves to pass right through them. Not so in the Netherlands.

Traveling west from The Hague, I reached the coast at the resort town of Scheveningen. There, I walked a paved trail along the crest of dunes that loomed high over the beach below; looking down, I saw a mother and young child slowly, slowly ascending a set of wooden stairs with more than one hundred steps. The dunes were covered with grasses, plants, and bushes of ankle to knee height, which performed the vital task of holding the sand in place against the ocean winds. And the dunes were enormously wide—a couple of hundred meters, I estimated. As I looked south toward Rotterdam and north toward Amsterdam, I saw no gaps in the line of protective dunes. But I later learned that six miles to the south, beyond my sightlines, trouble had been detected near the town of Monster. Those dunes, said Vellinga, "are only sixty meters wide and eroding on the sea side. So we are adding five million cubic meters of sand to make sure they don't erode away."

Coping with an additional three feet of sea level rise will mean that the Dutch coast "will need much more sediment in the future," Roelvink told me. "But we are lucky to have access to huge amounts of sediment in Doggers Bank [an area two hundred meters beneath the North Sea, halfway between Holland and the UK]. We also have 40 percent of the world's dredging fleet. So we can manage."

Phase two of the coastal adaptation plan, said Vellinga, "will develop alternative ways of dealing with coastal flooding, including reestablishing our system of secondary dikes." Secondary dikes are smaller earthen dikes that exist inland from the dunes and man-made barriers that comprise the first line of defense against the North Sea. The secondary dikes took shape over centuries as the Dutch pumped dry one polder after another; dikes were then erected around the dried polders to protect the farmland inside from flooding. "This [secondary] system was in good shape through the nineteenth century," said Vellinga. "But in the twentieth century we built railroads, roads, towns, and other infrastructure across these dikes, so the system was not as strong anymore. Now, we're preparing to reestablish this secondary system."

One day I drove northwest from Amsterdam with Vellinga's young colleague Laurens Bouwer to see what this meant in practice. In less than an hour we passed the town of Alkmaar, home to the national football champions the year before. Green meadows and a pretty patch of woods led to a small vacation town crowded with hotels and breakfast shops. The road then became a single lane. Soon we emerged from a second patch of woods to see before us, through gathering fog, a vast expanse of lowland that ended a couple of miles away in what looked like plump green hillsides: the back of the coastal dunes.

Before we approached for a closer look, Bouwer told me we had just entered the secondary dike system. In fact, the one-lane road we were riding on had bisected one of the secondary dikes. This dike was decidedly less impressive: about a meter tall, it was composed merely of raised earth and rendered all but useless by the road running through it. "These sleeper dikes, as we call them, are much older than the primary dikes along the sea," Bouwer said. "They were probably built in medieval times so that North Sea surges didn't end up going all the way to Amsterdam. Having a road go through the dike obviously makes it weaker. In an emergency, you could add sandbags to make it safer, but that is not a sufficient long-term solution. So we will upgrade."

We drove on as the fog and clouds started sprinkling rain. The road led to a parking lot on the landward side of the dunes. Pulling on rain gear, we walked up to an asphalt trail that extended along the top of the dunes. The wind was blowing strongly now; below us, the surface of the North Sea was flecked with whitecaps. "The dunes here are partly manmade, so we call them dikes, even though part of them is a natural sand dune," Bouwer said. "In the 1880s, they reinforced the dike with basalt blocks, but only in 1981 was this dike brought up to the standard [minimum] height of 11.5 meters [37 feet]. Nevertheless, the threat was still greater than realized. In the 1990s it was discovered that the strength of waves in the North Sea was one-third greater than thought. So in 2005 they added the concrete blocks you see down there at the foot of the dike. We place the blocks in irregular patterns, not straight lines, because that breaks up the waves better. But in the long run, this is not a good solution. The dike is fixed in its location, while the dunes are gradually retreating because of erosion, which raises the threat that the dike will separate from the dune."

"I can check the annual failure probability of this dike when I get back to the office," Bouwer said as we headed back to the car, "but for sure it is less than 1 in 10,000 years. That is the worst case, of course. But risk analysis must be based on the weakest link."

As the history of that dike illustrates, the Dutch bring centuries of investment, effort, and learning to bear in the race to prepare for climate change. One of their most impressive achievements was our next stop: the Afsluitdijk, a massive barrier dam that extends across thirty kilometers (twenty miles) of open water north of Amsterdam. North Sea storms and floods had claimed lives and destroyed property in this area until well into the twentieth century, but the Afsluitdijk put an end to such tragedies by preventing the sea from penetrating inland. Completed in 1932, the barrier transformed the body of water behind the dam—formerly a brackish bay, the Zuyder Zee—into a freshwater lake. A series of locks allowed ships and boats to pass through. In 1976, the road atop the dam was converted to a motorway; it was this motorway we now headed for.

On the way we drove past a smaller dike that featured a dozen sheep munching grass along its steeply pitched flanks. Still unlearned in the ways of dikes, I asked Bouwer if this grazing was such a good idea. "No, that's no problem," he said with a laugh. "In fact, the sheep are quite helpful, because they prevent the growth of trees along the dikes. Trees are not welcome because a storm could blow them over, which would weaken the dike's stability."

Soon we reached the barrier dam, and it was a sight to behold. It was 106 yards wide at the base. To our left, on the side facing the North Sea, it rose 25 feet above the waterline. When we stopped at a monument tower midway across, I looked down and saw more basalt blocks armoring the flank of the dam. "This dike is not as high as the one we saw a few minutes ago, but we don't see this as a problem," Bouwer said. "The waves that strike here are not as dangerous as the waves that come farther south, because there are barrier islands out in the sea that break up the waves before they get here." A bronze plaque on the south side of the barrier dam depicted three of the masons who had helped build the structure. Below them the inscription read, "A living nation builds for its future."

The Dutch approach coastal protection with great confidence but not arrogance. An inscription on one of the Delta Works dams in the southwest declares, "Here the tide is made by the moon, the wind, and us." But the natural elements — the moon and wind — outnumber the human, and these elements will grow stronger under climate change. Thus in the third phase of their flood protection upgrade, the Dutch "will decide where we should retreat — to give up land to the sea — and where we might advance," Vellinga said. "And we will advance in some places. For example, we plan to extend the port of Rotterdam farther into the North Sea."

But more common, perhaps, will be the fate of a recreation complex in the southern province of Zeeland. According to Pavel Kabat, a colleague of Vellinga's at Wageningen University and a leader of the Dutch adaptation effort, the national government originally intended to provide the low-lying area occupied by the complex with additional protection from strengthened dikes. But after three years of public discussion with local citizens, a "quite surprising" alternative emerged, said Kabat: the complex, which contained many hotels and high capital investments, would be demolished. Protecting the complex was deemed both too risky and too expensive; retreat — withdrawing human settlements from the vulnerable area — was the best option. "The owner of the complex was not happy about this idea," said Kabat, in what was doubtless quite an understatement. But like the farmers who were evicted from the polder Leusink showed me, the owner was compensated, in his case with majority ownership of a sea fish nursery that would be added in place of his complex.

Society's needs will continue to trump individuals' ambitions in the future, if the second Delta Commission gets its way. At the end of its 2008 report, the commission offered twelve recommendations. The second of them, trailing only the call for a tenfold improvement in flood protection levels, declared, "Any building in flood prone areas must be based on long-run cost-benefit analysis and costs must be borne by those who benefit, not by society as a whole."

The lesson of such episodes, said Kabat, is that we must change our way of thinking about climate change. Unwittingly echoing Ron Sims, he continued, "In the Netherlands, we no longer see [climate change] as a threat. We see it as an opportunity. We can't avoid it, so let's think about investing in solutions." Such solutions can only work, Leusink said, if they are first vetted in public discussions with all the relevant stakeholders. "Communication with the public about all forms of adaptation is essential," he told me. "The [national] government will do its part, formulating a plan, but then this plan must pass through a consultation process involving every level of government and every type of civic committee engaged with these issues. This is not for show; we really want their involvement. In fact, we cannot succeed otherwise."

"You Are More Vulnerable Than You Think"

It is true that the Netherlands enjoys certain advantages in the race to adapt to climate change. For one thing, it is rich, which makes it easier to make long-term investments and deploy advanced technologies. But Kabat and others emphasized that adaptation is only partially about deploying the right technologies; the cultural and sociological background of a country or community matters just as much.

Here the Netherlands boasts particularly great strengths. As noted earlier, centuries of history and practice have given the Dutch people a shared consciousness about the importance of water management, as well as methods for taking practical actions. The federal ministry of water resource management is powerful enough that it is considered a fourth branch of government alongside the executive, legal, and legislative branches. The water boards that decide local water issues (in conjunction with regional and federal bodies) date back to the 1100s, making them some of the oldest democratic institutions in the world. Dutch people respect the decisions these boards make, even when they disagree with them, which helps explain why the river widening near Nijmegen and the polder filling east of The Hague have gone forward rather than gotten mired in acrimonious protest and delay. Finally, centuries of joint effort to subdue the rivers and waves that threaten their homeland have given the Dutch a preference for acting by consensus in all spheres of public life, reaching decisions that all parties can live with, however reluctantly. Consensus does not always rule, of course, but it wins out often enough to give the Dutch an unmistakable sense of unity and cohesion, two values that promise to serve it well in the race against climate change.

A country's history and culture cannot simply be transplanted beyond its borders, but certain lessons from the Dutch experience are nevertheless useful for the rest of us.

The first piece of advice he would give to outsiders, said Vellinga, is to start now. "We don't have to know exactly what the world will look like beyond 2050," he told me. "It will be hard enough to prepare for 2050, and at that point we'll know more and can adjust our plans. Some people don't like this uncertainty. But the point is to start moving now, because it takes a long time to make these kinds of changes. When we built the dikes and sea barriers of the Delta Works, it took us thirty years to complete the project. Most of the delay was *not* on the technical side; the actual building of the structures took less than ten years. What took time was the *political* convincing needed to approve the decision and appropriate the funds and then the legal work needed to buy out people's property rights, rights that are well protected in most countries."

A second reason to start now, added Vellinga, is that even rich countries "are more vulnerable to climate change than they think, and the difficulties of adapting to it are bigger than they think. There is a general belief that the countries of the global North aren't likely to suffer from climate change because they are so rich and have so much technology and therefore can adapt to changes in climate. That is simply not true, because the costs of adaptation come before the benefits and you're talking about long-term risks. But the political systems of the North don't deal well with long-term risks: people are often reluctant to pay for protection against something that may not happen in their lifetimes."

Finally, bear in mind that adaptation is fundamentally a local activity. National and regional involvement is helpful, but real progress comes from "mobilizing local constituencies," said Vellinga. That's partly because the climatic conditions—and thus the necessary adaptation measures—will vary from locality to locality, so the day-to-day work of adaptation must be done locally, too. "In the end, you have to realize that nobody outside your local area is going to save you," said Vellinga. "It's up to you." Which means localities need solid information about the impacts they must prepare for, said Vellinga, adding that his Knowledge for Climate program has begun preparing such "climate atlases" for Dutch provinces (just as UKCIP has done in Britain). Localities must also find ways to pay for adaptation, preferably by linking it with other desired goods or services, as the New York City government did when it invoked the stresses expected from climate change as an additional reason to upgrade the city's Number 3 water tunnel.

"Beyond Two Meters, Even Dutch Engineers Get Worried"

Despite all this advice, don't think that the Dutch have all the answers about climate change. In fact, their performance is barely average on the essential challenge of mitigation. Rhetorically, they have long supported an aggressive approach, committing early to the Kyoto Protocol and consistently urging rich industrialized countries to adopt ambitious emissions reduction targets. But the Dutch have found it hard to reduce their own emissions. In interviews, current and former senior government officials differed on whether the Netherlands would achieve even the small reductions that are required under the Kyoto Protocol: 6 percent cuts from 1990 levels by 2012. None thought the country was on track to meet the tougher target the European Union has endorsed: 20 percent reduction from 1990 levels by 2020.

Herman Sips, a senior policy adviser at the Dutch Ministry of Housing, Spatial Planning and the Environment, acknowledged the shortcomings, which he blamed on the difficulty of regulating energy consumption, the lifeblood of modern society. Sips had helped formulate three of the Netherlands' Green Plans—national blueprints for integrating environmental concerns into larger national policies. The plans, which began in 1986, had achieved great success in many areas: reducing toxic emissions, improving habitat preservation, delivering cleaner soil and water. But on emissions cuts, Sips admitted, "clearly we have a long way to go." The conclusion he drew from the slow progress so far, he said, was that "gradual, small steps will never come to grips with this problem. You need big systems changes from the whole society: a common search for new technologies, and then organizing markets around those technologies to speed up their implementation."

"Personally, I don't think we'll meet our Kyoto commitment, though

I think it will be close," countered Hans Van Zist, a longtime government colleague of Sips's who now worked as a private consultant. Perhaps the biggest challenge, Van Zist said, was how to make mitigation strategies work within a growing economy: "Whatever we do to cut emissions from power plants and other single sources—and we've been fairly effective there—is being counteracted by the fact that there are more and more cars on the road, driving greater distances. And there is more aviation traffic as well. Schiphol airport is going to increase its volume by forty to fifty thousand flights a year and expand to six runways, which will make it a truly huge airport."

Echoing Sips's call for systemic reform, Van Zist added that meeting the challenge of mitigation allowed for "only one option: we have to decarbonize our economies." But instead the Dutch have been slow to develop solar, wind, and other alternative energy sources. "All the visionary documents on this are okay—I've coauthored some of them—but the proof is where the rubber meets the road," said Van Zist. Complaining that the government has often reduced subsidies for alternative energy when short-term goals such as the Kyoto reductions appeared to be in sight, he argued, "That means the big industrial companies stop moving toward the larger goals. If you really want to make the transition to a low-carbon economy, it's not about reaching a certain percentage by a certain date. It's about driving a comprehensive and continuing shift in overall economic behavior."

Van Zist doesn't let corporations and consumers off the hook either. "From a technical standpoint, we are in a position to make all cars much better environmentally, but we don't do it because it's not in the interest of the auto industry, which wants to make back its money from the old technologies first. As long as it takes so long to move to the next level of technology, we're fooling ourselves to think we are coping with this problem." Nor were most individuals willing to change their behavior. "I've conducted lots of focus groups and surveys on this issue, and people in the Netherlands say they are very proud of what they are doing for the environment. Why? Because they are separating their household waste for recycling! Recycling is important, of course, but it is just a small step toward the changes that we need to make. If you ask people about driving less, they will use everything in their power to justify using the car. In my time, you walked to school. Now, parents drive you, and they say they're doing it because it's too dangerous to walk. Why is it too dangerous? Because of all the cars! It's crazy."

The irony of all this, of course, is that each day of delay in cutting emissions puts Dutch people at greater risk of flooding and other impacts that could overwhelm even their nation's extraordinary defenses. Which has made me think twice about Chiara relocating to the Netherlands someday. If global emissions aren't reversed soon, very few places on earth will be safe. Skillful adaptation can bolster a given location's defenses in the short to medium term, but there are limits to what even the Dutch can achieve against unimpeded climate change. Vellinga has urged the government to increase the safety levels required against North Sea storms from the already formidable level of 1 in 10,000 years to a level of 1 in 100,000 years. Safety levels for meltdowns of nuclear reactors are generally set at 1 in 1 million years, he noted, "so for a low-lying country such as the Netherlands it is not so strange to go for a 1-in-100,000 chance per year." And having already agreed to prepare for 1.5 meters (5 feet) by 2100, Vellinga and his colleagues have already begun thinking about what happens if seas continue rising after that. "Up to 2 meters [6.5 feet] of sea level rise, we believe we can do the job," he told me. "Beyond that, even Dutch engineers get a bit worried."

6 Do You Know What It Means to Miss New Orleans?

Hurricane Katrina was a global warming wake-up call, but it remains to be seen how awake the patient is.

—MARK DAVIS, Senior Research Fellow, Tulane University Law School

THE DUTCH MAY be the world's leaders in coastal protection today, but one hundred years ago they were taking lessons from, of all places, New Orleans. In 1913, a Tulane University engineering graduate named A. Baldwin Wood invented a new type of water pump. Relying on suction rather than lifting to pump water, the pump doubled New Orleans's capacity to drain low-lying areas after heavy rains. The so-called Wood pump became world-famous, report Pulitzer Prize–winning journalists John McQuaid and Mark Schleifstein in their book *Path of Destruction*. When the Dutch found that their own water pumps were inadequate for reclaiming submerged land, they sent for Wood. But like countless New Orleans locals then and now, Wood did not want to leave his hometown. So the Dutch came to him, sailing across the Atlantic to consult with the engineer in his beloved New Orleans.

A century later, the roles were reversed. In the aftermath of Hurricane Katrina, numerous Dutch water experts came to New Orleans to advise the stricken city on how to drain the city and fashion better defenses against future storms. More than one of the Dutchmen were surprised by what awaited them—by the shoddy design and disrepair of the local levee system and the secretive, disorganized decision-making process that governed it.

"It's very confusing here," Bas Jonkman, an adviser to the Dutch ministry of water management, told me over coffee one afternoon in the French Quarter. "Everyone is making their own plan: the Army Corps of Engineers, the state of Louisiana, the city of New Orleans. One lesson we have learned in the Netherlands is that from the beginning you have to include all stakeholders in the process, and it must be a transparent process. Otherwise, everyone defends their own plan and attacks the others instead of working together to find the best plan."

Also baffling was the Americans' insistence on trying to protect everyone, everywhere. The Army Corps of Engineers, for example, was proposing to erect a massive new system of armored levees across sparsely populated southern Louisiana. Nicknamed "Morganza to the Gulf," the proposed levees would extend thirty to seventy miles (depending on which option was selected) across terrain that was mainly wetlands. "This idea doesn't make sense," said Jonkman. "By blocking the flow of river and tidal water, such levees would end up ruining the wetlands, which provide valuable protection against hurricanes' storm surges. Much better would be to accept that the outlying communities are impractical to defend and pay to relocate them farther inland. That's another lesson from the Netherlands: society must recognize that there will be losers in such situations, and the losers must be fairly compensated."

But it seemed futile to suggest such alternatives to the Corps. "There's a culture [within the Corps] where you can't have open discussion," said Jonkman. "They get very defensive."

I interviewed Jonkman in March 2007, eighteen months after Hurricane Katrina had sent New Orleans on a globally televised journey to hell. It was my second visit to the city since the storm, and it was still in shambles. I spent the better part of two days driving back and forth across the hardest-hit parts of the city: the infamous Lower Ninth Ward but also New Orleans East, the much larger community just to the north.

I also toured the neighborhoods of Gentilly, Lakeview, and St. Bernard Parish. Traveling up one street and down the next for hours on end, I drove past variation after variation on a single theme: upended cars and boats, wrecked houses with roofs smashed in, front walls spray-painted with numbers and dates signifying when the house had been searched and how many dead bodies had been found. Television could not convey the enormity of the devastation, for it went on not merely block after block but mile after mile, especially in New Orleans East and St. Bernard Parish. Eighteen months after Katrina, most of New Orleans still looked like a ghost town—mostly empty, houses dark, streets deserted.

The ruination of New Orleans offers a lesson—a warning, really—about what can happen to people and places that fail to prepare for the impacts of climate change. At the moment, that includes most people and places on earth. At a time when weather-related disasters are projected to increase in severity and perhaps frequency, most people are waiting until disaster strikes before putting proper safeguards in place.

We cannot say yet exactly how much responsibility global warming may have borne for Hurricane Katrina. The climate system is complex, and scientists have not yet compiled enough historical data on hurricanes to perform the kind of analysis that detected the climate signal within the European heat wave of 2003. Nevertheless, Katrina certainly fit the pattern. Even in 2005 many scientists agreed (and support has only solidified since then) that global warming has made extra-strong hurricanes more likely because it encourages hot oceans, a precondition of hurricane formation. "It's a bit like saying, 'My grandmother died of lung cancer, and she smoked for the last twenty years of life—smoking killed her,'" explained Kerry Emanuel, a professor at the Massachusetts Institute of Technology who had studied hurricanes for twenty years. "Well, the problem is, there are an awful lot of people who die of lung cancer who never smoked. There are a lot of people who smoked all their lives and die of something else. So all you can say, even [though] the evidence statistically is clear connecting lung cancer to smoking, is that [the grandmother] upped her probability." In the same way, concluded Emanuel, humans are "loading the climatic dice in favor of more powerful hurricanes in the future."

Remember, scientists have calculated that the world as a whole will experience at least four additional mega-hurricanes a year by 2050; that is, the frequency of Category 4 and Category 5 hurricanes will increase from thirteen a year today to seventeen by mid-century. That means, in the words of Joseph Romm, a climate scientist at the Center for American Progress, that by 2050 there will be four more hurricanes each year that are big enough to demolish entire cities. In the United States, the locations most at risk include most of the Gulf coast from Texas to Florida and most of the Atlantic coast from Florida to New York. Internationally, the danger zone centers on the tropics of Asia: India, Bangladesh, Indonesia, the Philippines, Thailand, Vietnam, China, and much of Japan. The Caribbean and southeastern Africa are also at risk.

Hurricane Katrina ranks as a defining event of global warming's second era partly because it made these risks explicit; the televised destruction of New Orleans put people and institutions around the world on notice that something similar could happen to them. As a matter of science, the heat wave of 2003 was an earlier example of the second era's decisive characteristic: the arrival of actual climate impacts. But Hurricane Katrina had a much more powerful, and more global, effect on public awareness than the 2003 heat wave did. Because TV cameras managed to enter New Orleans—even when government emergency vehicles somehow could not—the outside world saw it all: the hurricane's westward journey across the Gulf of Mexico before turning north toward New Orleans; its striking the Gulf coast on Monday, August 29, 2005; the sheer physical power of the storm; the failure of levees that left 80 percent of the city underwater; the horrifying circumstances local people soon found themselves in; the inept response of American government at all levels—local, state, federal. These circumstances were made all the more shocking by the fact that the victims lived in the richest country in the world. As the drama continued, each day seemed to bring fresh examples of human suffering and official incompetence. The rest of the world watched this drama unfold live, in real time, like a macabre reality show nobody could turn off. The effect was to sear Katrina into the mass consciousness, in America and throughout the world, in ways few would forget.

Clearly, New Orleans was woefully unprepared for Hurricane Katrina. But what the TV coverage failed to convey—what is still not appreciated today, more than five years later—is that such unpreparedness is by no means unique. I found much the same in many places I visited for this book: on the other side of the Gulf of Mexico, in Tampa Bay, Florida; on the other side of the country, in California's capital of Sacramento; on the other side of the world, in China's commercial capital, Shanghai. Robert Bea, a professor at the University of California, Berkeley, and a coauthor of a landmark National Science Foundation report on Katrina, identified three major U.S. cities—Miami, Houston, and Washington, DC—that are no better protected today than New Orleans was before Katrina. "In 2006 I went to Houston and found the very same problems we uncovered in New Orleans," Bea told me. "Just south of the city were levees built out of sand, I-walls instead of [stronger] T-walls, walls built shorter than proper design required—replicas of what went wrong in New Orleans. If those levees [near Houston] were to fail, you'd take out two-thirds of the U.S. oil refinery capacity."

If we hope to do a better job of protecting these and countless other ill-prepared communities in the future, it is crucial that the right lessons be drawn from the Katrina tragedy. We must understand not only why New Orleans was left so vulnerable in the first place but, equally important, why the effort to resuscitate and protect the city after the disaster has been such a misguided, ineffectual mess. Every locality is different; experts emphasize that there is no one-size-fits-all model of climate change adaptation. Nevertheless, if we compare the failures in New Orleans with the successes in the Netherlands, one lesson stands out: social context matters more than technological prowess. The Dutch have been relatively good at preparing for climate change largely because of their long history of consensus-based water management and their shared belief in social planning. By contrast, Louisiana's efforts have been crippled by the state's history of poor government, its dysfunctional relationship with the Army Corps of Engineers, the power of its oil and gas interests, its continuing reluctance—even after Katrina—to acknowledge the reality of global warming for fear that might harm oil and gas production, and an abhorrence of taxes and public planning as somehow socialistic. (Only after Katrina did Louisiana adopt a statewide building code.)

Whether Louisiana can adapt to climate change will depend on whether the prevailing mindset within the state, the social context, can change quickly enough, said Mark Davis, a senior research fellow at Tulane University Law School and the former director of the nonprofit Coalition to Restore Coastal Louisiana. "There is still pressure for development, for sprawl, for the idea that property is king and that everyone, no matter where they live, has a right to the same level of protection—that it's almost antidemocratic not to protect everyone," Davis said. "All that gets in the way of making the hard choices that are inevitable."

Dirty Practices

Rich countries "are more vulnerable to climate change than they think." When Vellinga, the Dutch climate scientist, offered that comment about other nations' vulnerability to climate change impacts, he was referencing the aftermath of Hurricane Katrina. "Safety levels around the world are usually set by law at a 1-in-100-years level of protection," he told me. "But in reality, the level is often closer to 1 in 20 years. This is because of what we call 'dirty practice'—gates are left open, levees aren't well maintained. You saw this in particular in New Orleans."

Often the "dirty practices" that left New Orleans poorly defended had been going on for decades. Levees were poorly designed and shoddily built by an unholy alliance between the Army Corps of Engineers and local construction companies that put bureaucratic habit and private profit above public safety. Levees were lackadaisically maintained because of municipal flood boards whose members cared more about lavish lunches than rigorous inspections. The result was a flood protection system with so many flaws that it qualified as "a system in name only," as the Corps later admitted.

The federal and state governments further undermined hurricane defenses by helping commercial interests to destroy the coastal wetlands and cypress swamps of southern Louisiana. To guard against flooding but also to aid the shipping industry, the Corps over the past eighty years encased the Mississippi River in levees, preventing its silt from replenishing the wetlands. To encourage oil and gas production off the Louisiana coast, state and federal regulators allowed industry to build thousands of miles of navigation channels through the wetlands. Scientists and environmentalists warned at the time that all this invited disaster. Wetlands and swamps offer unsurpassed protection against hurricanes, for their grasses and foliage act like speed bumps, weakening the force of incoming storm surges. But profits took precedence over ecology, and the state of Louisiana ended up losing 1,900 square miles of wetlands, an area the size of Delaware, between 1932 and 2000.

Meanwhile, in New Orleans, city officials never bothered to develop sound procedures for evacuating people before a hurricane or for getting aid afterward to those too sick or poor to leave. Thus, less than forty-eight hours before Katrina made landfall, with hurricane watchers warning that this was "the big one" they had long feared, Mayor Ray Nagin declined to order a mandatory evacuation of the city. Why? Because, Nagin later said, he was unsure whether he had the legal authority to do so. Nor were federal agencies any better prepared. With bloated corpses floating in the streets and desperation mounting throughout the city, the Federal Emergency Management Agency proved unable to deliver water, food, and medical aid for days after the storm, turning FEMA into a local curse word (as in "I got FEMA'd").

These failures were all the more inexcusable given that authorities had received many high-profile, quite specific warnings over the years that New Orleans was at grave risk of a disaster like Katrina. Indeed, Hurricane Betsy in 1965 hammered the city so hard—putting 20 percent of it underwater and killing seventy-five people—that the federal government ordered the construction of a levee system, the very system that later failed during Katrina. As far back as 1981, the U.S. National Hurricane Center director, Neil Frank, was telling the national media that New Orleans was "one of the most vulnerable places in the U.S." and had no real "workable evacuation plan. . . ." In June 2002, the hometown newspaper, the *Times-Picayune,* published a series of investigative articles by Schleifstein and McQuaid that concluded that tens of thousands of people could die from a large hurricane and that stronger levees and revitalized wetlands were essential protections. The series won the two journalists a Pulitzer Prize, and similar warnings were delivered at about the same time by other media, including *National Geographic.* Nevertheless, no serious action was taken.

After Katrina, federal lawmakers finally focused on the physical vulnerability of New Orleans, though some now cited that vulnerability as a reason not to rebuild the city. Republican Dennis Hastert, the Speaker of the U.S. House of Representatives at the time, opened the debate when he questioned if it was worth spending billions of taxpayer dol-

lars to defend a city as low-lying as New Orleans. I heard similar comments many times while reporting this book—always, I noticed, from
people who lived far away. New Orleans residents—no surprise—saw
the issue differently.

Locals knew what few outsiders recognized: New Orleans was essential to the national economy. "Purely on an economic basis, the nation
needs a port at the mouth of its largest river," said John Barry, the author of the classic book on the 1927 Mississippi River flood, *Rising Tide.*
After Katrina, Barry co-chaired a commission charged with planning
New Orleans's future; in that role, he often had to explain to Washington lawmakers and other distant power brokers that New Orleans was
the busiest port in the world and the main gateway through which the
crops of the Midwest reached foreign markets. Louisiana's oil rigs and
refineries supplied a quarter of America's petroleum. Its fishermen provided a third of the nation's seafood. "Once you've paid to protect [the
economic value of New Orleans]," Barry added, "protecting the people
who live there is almost a throwaway cost."

Of course, abandoning New Orleans would also mean abandoning
an irreplaceable jewel of America's history and culture. Founded in 1718
by French Canadian explorer Jean Baptiste Le Moyne, Sieur de Bienville,
New Orleans is in fact older than the United States. Its riverside market
had been a headquarters of the transatlantic slave trade that shaped so
much of U.S. history; its musicians gave birth to jazz, America's most
original art form.

Most infuriating about outsiders' reluctance to help rebuild New Orleans, locals said, was that it was based on misinformation. "It wasn't
Hurricane Katrina that put 80 percent of New Orleans underwater, it
was the inexcusable failure of our levees," said Sandy Rosenthal, a city
resident who cofounded the citizens group Levees.org. "Our city is not
as far below sea level as most of the Netherlands is, but the Dutch do a
fine job of protecting themselves. If they can do it, why can't we?"

"Most people still don't know how and why New Orleans flooded,"
observed Hassan Mashriqui, a professor at the Louisiana State University Hurricane Center who studied the storm and its aftermath exten-
sively. "I remember getting a phone call from a producer at NBC News
a few months after Katrina, when many residents were asking to move
back to their old neighborhoods. The producer kept asking me, 'Why

would anyone want to live in the Lower Ninth Ward? It's so low there.' I told him the Lower Ninth was four feet higher than Lakeview [a middle-class neighborhood three miles away that also suffered severe flooding]. He wouldn't believe it. He kept asking, 'Are you sure of that? Are you sure?'" Mashriqui continued, "Katrina showed me as an engineer that elevation had nothing to do with who got flooded and who didn't. Flooding was a function of where levees held and where they didn't, and in many places they didn't."

Locals blamed the levee failures on the Army Corps of Engineers, which had, after all, supervised their design and construction. The Corps blamed Mother Nature: the problem, officials told the media, was not the quality of the levees but the strength of the hurricane. Congress, Corps officials added, had ordered the Corps to provide New Orleans with only Category 3–level protection, but Katrina was a Category 5 hurricane. "Sure, it hurts to hear so much criticism from locals, including our family and friends, but you just chalk it up to ignorance," said Troy Constance, a New Orleans native who was chief of the Restoration Branch of the Office of Coastal Protection and Restoration of the New Orleans District of the Corps. Constance told me during a tour of the reconstructed Industrial Canal that other locals simply "don't know what happened. We got hit with an excessive event that no one anticipated, a Category 5 hurricane."

But the factual record does not support the Corps. Although Katrina was temporarily a Category 5 hurricane during its journey across the Gulf of Mexico, by the time it made landfall it registered as barely a Category 3 storm, according to the federal government's National Hurricane Center, the recognized authority on such matters. What's more, Katrina had swerved sharply just before making landfall, thus sparing New Orleans from a direct hit. The documentary record points instead to errors at the hands of man, in particular the Corps. An investigation conducted by Mashriqui's LSU Hurricane Center colleague Ivor van Heerden revealed, among other things, that the notorious Seventeenth Street levee had *not* been "overtopped" by flooding, as the Corps had implied. Like five other levees in and around New Orleans, the Seventeenth Street levee had suffered a catastrophic failure—a collapse, in common language—caused by improper design and poor construction. As chronicled later by the National Science Foundation, the errors that compromised New Orleans's flood defenses were basic. Levees were constructed on sandy soils. Support pilings were not driven deeply enough into the earth. I-walls rather than sturdier T-walls were used for the levees' sides. In other words, had the levees been designed and constructed properly, New Orleans might have emerged relatively unscathed from the Category 3 force of Hurricane Katrina.

Instead, the floodwaters changed the very color and character of New Orleans, perhaps forever. Although both white and black and rich and poor suffered grievously after Katrina, poor and nonwhite residents ended up losing a disproportionate share of their homes. "The city has been depopulated of poor and working-class black people, and it has been made difficult for middle- and mid-upper-class blacks to return," Beverly Wright, the director of the Deep South Center for Environmental Justice at Dillard University, told me in 2008. Wright noted that plenty of private and public money had flowed into New Orleans after Katrina, but, she said, it had been channeled largely to white areas such as Metairie and Lakeview, while black neighborhoods that were hit just as hard, such as New Orleans East, where she lived, had gotten virtually no additional flood protection. Meanwhile, low-income housing was being torn down and replaced with so-called mixed-use housing that was mainly expensive apartments. "The lesson is that if you are white and wealthy, you'll get everything the government has to offer," said Wright. "If you're black, whether you're rich or poor, you'll get nothing."

The most hopeful things done in New Orleans after Hurricane Katrina came not from government or the private sector but from civil society—from church and school groups, university experts, and nongovernmental organizations whose motivation was not financial but moral. This was true in the immediate aftermath of the storm, when hundreds and perhaps thousands of individuals descended on New Orleans to rescue people off of rooftops and provide them with food, water, and shelter, and it remained true thereafter as the city struggled to get back on its feet. While government at all levels was merely talking about providing recovery assistance, countless high school and college kids, church groups, and retirees were paying their own way to New

Orleans and helping residents to gut and begin rebuilding their ruined houses. Local people also rose to the occasion, often undertaking leadership roles that surprised even themselves.

"My parents were both members of the civil rights movement back in the day, and I marched and boycotted with them, but I left activism after I got married and had kids," said Patricia Jones of the Lower Ninth Ward Neighborhood Empowerment Network Association, one of the many local efforts that sprang up after Katrina. After chairing a meeting of the group, Jones, wearing an "I Am a Survivor" T-shirt, told me that the hurricane had made her return to activism unavoidable. It had also convinced her of what the local Sierra Club activist Darryl Malek-Wiley had long been suggesting: renewable energy is essential to a healthy future for New Orleans. "Whatever money does come into the Lower Ninth, we want it to go to community-led green development," Jones told me. "We want to build up our own wind turbines, energy efficiency, and geothermal heating and cooling systems, which will also create jobs for local people. Green is the only kind of future that makes sense here."

"We're Repeating the Same Mistakes"

Going forward, the question is whether New Orleans can realistically be defended against the Category 4 and 5 hurricanes that will become more common during global warming's second era. The Dutch example suggests that, technologically, the answer is yes. The social context of New Orleans, however, gives much less reason for confidence.

"It's very important for the rest of America to understand that we can protect Louisiana if we want to," said van Heerden, who, in his book *The Storm*, urged a three-layered approach to hurricane protection known as "defense in depth." "For your inner layer of defense," van Heerden told me, "you put hardened levees or flood walls in front of major population centers [such as New Orleans] or other high-value assets. You protect that inner layer with a middle layer of defense, which is comprised of as large an expanse of swamp or wetlands as possible to absorb and weaken incoming storm surges. The data suggest that every mile of wetlands reduces storm surge by 0.7 feet, and every mile of swamp reduces it by 5 to 6 feet. Finally, you protect that middle layer with a third layer—barrier islands out in the ocean proper, which also absorb and weaken storm surges." To restore the inner layer of defense, van Heerden suggested that the Army Corps of Engineers be stripped of its monopoly over levee construction; instead, he suggested, let a design competition be held, "like they do in the Netherlands, to pick the best possible approach." To rebuild the wetlands and swamps that would form the middle layer of defense, he advocated letting the Mississippi and other rivers run free in places so they can disperse their silt and raise land elevation. The same tactic would also nurture the barrier islands that would serve as the outer layer of defense.

Van Heerden's strategy would protect 85 percent of the people, towns, and infrastructure of coastal Louisiana, he said, and at reasonable cost: $15 billion, compared to the $200 billion in economic damages caused by Katrina. But he conceded that his plan had a key drawback: it left 15 percent of the state—mainly southern coastal communities—outside the defenses. To extend protection to the entire state would require such extensive levees that vast areas of wetlands would be destroyed, undermining the crucial second layer of defense. It would also vastly increase the cost. If even the most vulnerable sections of Louisiana's coast were included, and the Army Corps of Engineers was left in charge, van Heerden said, the cost would rise to an estimated $100 billion. "We have to be honest with people—we can't save everyone," he told me, adding that, as in the Netherlands, fair compensation should be paid to those who are left outside the defenses.

As sensible as this approach sounds in theory, the prevailing social context has made it very difficult to put into practice. The Netherlands can implement retreat-with-compensation largely because the nation's laws, values, and history support the idea. But retreat-with-compensation is a much harder sell in the United States, especially in the conservative South, where individualism is treasured, private property rights are sacred, and government is despised except when it is subsidizing oil and gas production.

For example, retreat-with-compensation is all but incomprehensible to the Army Corps of Engineers, and the Corps—amazingly—remains in charge of post-Katrina rebuilding. I say amazingly because, well, Katrina is hardly the first stain on the Corps' reputation. The fact is, the Corps has long had a terrible track record throughout the United

States, as journalist Michael Grunwald documented in blistering detail in his book *The Swamp*. Time and again, flood defense and navigation projects overseen by the Corps have come in wildly over budget even as they wrecked ecosystems and failed to deliver the economic benefits that were promised. Yet the Corps was rarely if ever called on its behavior, for it enjoyed a congressionally bestowed immunity to lawsuits regarding flood control projects.*

I asked many experts why the Corps was allowed to retain control of post-Katrina reconstruction but never got a good answer. The closest thing to it came from Robert Bea of UC Berkeley, who had often worked with the Corps during his long career in private industry but was now a frequent critic. But even Bea said the nation has no choice: only the Corps possesses the technical and financial resources needed to handle levee construction and other big infrastructure projects. Still, if it is not possible to fire the Corps, it deserves a thorough overhaul, the rescinding of its immunity to lawsuits, and—a suggestion from Bea—a requirement to submit its projects to peer review. Neither political party in Washington has shown any enthusiasm for such reforms, however, despite the Katrina debacle. Thus the Corps lumbers on, a clear liability to America's ability to construct the flood defenses demanded by the second era of global warming.

The Corps, of course, does not see it that way. Major General Don Riley, who as director of the Corps' civil works division oversaw rebuilding in Louisiana, told me that the Corps had learned lessons from Katrina. Moreover, the Corps pledged to do "whatever was necessary" to set things right, including being more open to outside experts and pub-

---

*Hurricane Katrina, however, gave rise to a stunning exception to the Corps' immunity. On November 18, 2009, a federal judge ruled that "the negligence of the corps" had contributed to the failure of a navigation channel that caused much of the flooding in the Lower Ninth Ward and St. Bernard Parish. Responding to a lawsuit filed by New Orleans homeowners, the ruling represented the first time the federal government was held liable for damages from Hurricane Katrina. The key to victory, Robert Bea later explained, was the plaintiffs' contention that the channel in question, the Mississippi River Gulf Outlet, was not a flood control project but a navigation waterway and thus the Corps' legal immunity was not applicable. The government said it would appeal the ruling, which could force the government to pay out tens of millions of dollars, or more, to homeowners whose property was damaged.

lic feedback. But Riley pointed out that the Corps takes its orders from Congress, and he did not see how to deliver the Category 5 protection Congress had demanded without adding an extensive system of "structural defenses"—in other words, levees—across southern Louisiana. I noted that critics feared that this would doom vast swaths of wetlands, undermining the larger goal of hurricane defense. Why not move the line of levees inland, as critics had proposed? That was possible in theory, Riley replied, "but then you leave people outside the ring of protection," which was not what Congress had in mind.

Louisiana state officials also worried early on that the Corps would shortchange wetlands protection in favor of its traditional preference for large levees, and that is eventually what happened. "We're not going to let [the Corps] go down that road," Robert Twilley, the chief scientific adviser to the state's planners, told me in 2007. "If we don't restore our wetlands, the levees won't last and neither will our economy." A year later, Twilley told me the state and the Corps were still haggling, with the Corps continuing to favor "the sort of brute-force engineering" of extensive levees while the state was seeking greater wetlands restoration. In 2009, the Corps finally issued its plan for protecting Louisiana from future hurricanes, and the state of Louisiana was not pleased. Forsaking the bland language usually found in government statements, the Louisiana Coastal Protection and Restoration Authority, which answers directly to the governor, said state officials were "very frustrated and disappointed" that the Corps' plan lacked "recommendations to protect and restore coastal Louisiana despite multiple laws passed by Congress requiring those recommendations." The statement further complained that state officials had been "left out of the writing of the plan despite assurances from the Corps of Engineers that the state would be a partner in the process."

"We're repeating the same mistakes that got us into this mess," said Oliver Houck, a professor of law at Tulane University in New Orleans whose decades of studying coastal policy in Louisiana resulted in a brilliant analysis of Katrina and potential next steps titled *Can We Save New Orleans?* Houck believed that New Orleans could still be saved, but it would require dramatically different policies from those of the past, as well as keeping future sea level rise to a maximum of three feet. "A lot of lessons have been learned since Katrina," Houck told me, "but those lessons can't be applied because the politics of the situation haven't changed, and it's the politics that drive this. You have to realize that people down here look at long-term planning, especially by the federal government, as communism."

For decades, oil and gas had been the heart of the Louisiana economy and a major source of the state government's revenues. This made politicians, the business community, and much of the public reluctant to accept that global warming was really a problem. Since the start of the first era of global warming in 1988, none of the state's governors or members of Congress, whether Democrat or Republican, had spoken out or voted as if climate change was a threat, even though Louisiana was considerably more at risk than most of the fifty states in the Union. As late as 2010, despite having seen what Katrina did to her state, Democratic senator Mary Landrieu was seeking to strip the U.S. Environmental Protection Agency of its authority to regulate greenhouse gas emissions. Republican governor Bobby Jindal was urging defeat of President Obama's climate legislation. A state that so forcefully resists the mitigation of climate change can hardly expect to be successful in adapting to it. As Dutch adaptation expert Richard Klein observed earlier in this book about the Bush administration's foot-dragging, "You can't adapt to a problem you don't admit exists."

"We're also still relying on hard structure strategies [i.e., levees] that are proven failures," Houck added. "But levees are what the Corps knows, so it's their answer to everything. Everyone agrees you need some levees, but [the Corps is] doing it very aggressively. It would be much better to move the levees twenty miles back [from the coast] and put ringlets [of levees] around the outer communities you want to protect. But the Corps won't do that because of the politics at play. See, building levees allows housing and commercial development to proceed behind the levees, and the Corps is tied at the hip with the construction and real estate interests that benefit from that."

By April 2010, when I last spoke with Houck, the Corps was pushing an even more ambitious levee program. "They want to build a dike across the eastern Gulf of Mexico, levees from Morgan City [near Louisiana's southern coast] to New Orleans, and new levees from New Orleans over to Mississippi," he said. "These levees would cut off coastal wetlands from upland fresh water and open 400,000 acres of wetlands to development. The levees are projected to rise fifty-three feet in the air and rest on soils that are sinking. It's madness, but madness that makes big money for contractors and real estate developers and spares local officials from making very hard, if obvious, land use decisions."

Down the hall from Houck, his colleague Mark Davis observed that the effects of these Louisiana political and cultural habits were compounded by a basic trait of human nature. When faced with unsavory choices, humans have a tendency to put a decision off until tomorrow and hope that tomorrow doesn't come. But "that's the final lesson of Katrina," said Davis. "Tomorrow does come. And you're not going to like it if you didn't do everything you could to prepare for it."

Florida Bets Against Climate Change

My first impression of Florida's Tampa Bay area was that it was much better prepared for hurricanes than New Orleans had been. Like New Orleans, the Tampa Bay metropolitan area faces immense inherent challenges. For one thing, it contains the most densely populated county in Florida, Pinellas County. What's more, its geography—much of Pinellas County occupies a peninsula, cut off from the mainland by the bay—makes evacuation extremely difficult. "Three of our main evacuation routes go over water," said Gary Vickers, a soft-spoken, balding redhead who ran the Emergency Management Office in Pinellas County. "So they might well be cut off in a storm."

But Vickers was plainly a competent man who took the risks of hurricanes seriously; as he reviewed the procedures he and his colleagues follow every day during hurricane season (which lasts from June 1 to November 15 in Florida), it was hard not to be impressed by his thoroughness and dedication to duty. On the eastern side of the bay, in the city of Tampa, Mayor Pam Iorio told me she and her staff had begun reviewing their hurricane preparedness a year *before* Katrina. Their motivation? The extraordinary 2004 hurricane season, when Florida was hit by five major hurricanes, three of which were Category 3 or stronger. Tampa had gotten a particularly bad scare from Hurricane Charley, a Category 4 storm that was heading directly toward the bay before it turned at the last minute and battered Charlotte Harbor, a town some sixty miles to the south. Hurricane Katrina a year later "was a major wake-up call," Iorio continued, "and we made a number of changes in disaster planning." The emergency operations center was moved to higher ground. Regular dialogues were begun among local and state agencies. Satellite telephones were bought so emergency staff would not lose communications if trunk and cell phone infrastructure was knocked out.

Vickers and his counterpart, Larry Gispert, the director of emergency operations for Hillsborough County, which contains the city of Tampa, were adamant that they would never fumble pre-storm evacuation or post-storm relief efforts the way authorities had in New Orleans. "[Mayor] Nagin was asking the day before Katrina hit whether he had the legal authority to order a mandatory evacuation. Sor-ry," bellowed Gispert, a big, blustery man who obviously took his job very seriously. "That question should have been resolved in everyone's mind long before. Here, we have specific procedures that bring together all the relevant agencies under clear lines of authority, and once the decision [to evacuate] is made, we execute." The Tampa metro area was divided into evacuation zones—A, B, C, D, and E—depending on a given zone's vulnerability to flooding and its options for evacuation; thus, authorities could order the most vulnerable places evacuated first. Gispert estimated that about 10 percent of the population was too sick, elderly, or otherwise disadvantaged to leave on their own. "We have a very sophisticated plan to help those people," he said. "We've got a fleet of specially dedicated buses and all you have to do is go to your local bus stop; the bus will be by within twenty-five minutes and take you to a shelter."

But as I dug deeper, I came to believe that the Tampa Bay area, and Florida in general, confirmed the truth of Pier Vellinga's warning: you are more vulnerable than you think. The likelihood of more Category 4 and 5 hurricanes in the years ahead is worrisome news for Florida for two reasons: first, current evacuation plans, though impressive on paper, are ignored by much of the population, who decide to "ride out" storms in their homes; and second, almost none of the buildings in the state are capable of withstanding more than Category 3 hurricanes. Most at risk are the hundreds of thousands of mobile homes in Florida—there are fifty thousand of them in Pinellas County alone—which a Category 4 or 5 storm could "hurl through the air like missiles," said Vickers. Nevertheless—and this is where social context again rears its head—even disaster officials as dedicated as Vickers and Gispert were not calling for the state to upgrade its building codes to require Category 4 and 5 levels of protection. Why not? Because, they said, it would sink Florida's economy.

Echoing virtually every disaster official I've interviewed anywhere, Gispert and Vickers said that many people simply refuse to evacuate, even when weather forecasts, emergency officials, and common sense all say it's time to go. When Hurricane Charley was bearing down on Tampa Bay and authorities ordered people out, "430,000 people should have evacuated, and we anticipated that 108,000 would actually do so," Vickers recalled. "But our surveys after the storm indicated that less than 10,000 people ended up leaving."

Evacuations inconvenience people and cost businesses money, so many resist or resent them, especially after episodes like Hurricane Charley, when the threat to Tampa Bay didn't actually materialize. By contrast, the commanders at MacDill Air Force Base, which is located on a peninsula that juts deep into Tampa Bay, do not hesitate to evacuate it when necessary, said Larry Clark, the base's head of the Office of Emergency Management. "But they don't have to worry about the politics of evacuations," Clark added. "Pinellas County ordered an evacuation on July Fourth weekend that led to millions of dollars of tourism money being lost, and there were lots of complaints from business people after that, believe me."

In Key West, Florida, tourists are evacuated thirty-six hours before the arrival of even a Category 1 storm, partly because the only route out of town is a two-lane highway that stretches over forty-two bridges and 100 miles before reaching the mainland. Under the circumstances, "even one accident means gridlock," said Irene Toner, the director of emergency services for Broward County. Yet even after Katrina, most locals ignored evacuation orders, said Toner, adding, "People here are very blasé. They'll tell you, 'My grandmother lived through plenty of hurricanes. We can ride it out.' They just don't realize what a big storm would do. We will be cut off here from water, power, sanitation, medical care. Life is going to be very hard. So why, why, why stay behind and put your family in that position? But people just don't get it."

Of course, the pledge to "ride out" a hurricane implicitly assumes that one occupies an adequate shelter. Florida law requires all buildings to be resilient to wind speeds that in most places are equivalent to Category 3 hurricanes, but in reality many are not, said Vickers. Upgrades tend to be made when a building changes ownership; Vickers estimated that full compliance was still ten years away. And Category 3 protection will not do much good against a Category 5 hurricane. Even a Category 3 or 4 storm, Vickers said, "would knock down or ruin most buildings in Evacuation Zones A and B. We'd lose almost 100 percent of our mobile homes." In the city of Tampa, the oldest public hospital sits on the edge of the inner harbor, with nothing to shield it from the path of a hurricane's storm surge. Evacuating patients would be impossible, said hospital spokesman John Gunn. Instead, he told me, the plan was to ride out any hurricane. But the building was only Category 3 resilient, so how exactly would that work in the case of bigger storms?

"Economically, we can't afford to build to a Category 4 or 5 level," said Gispert, "much as I as an emergency professional would like to see that happen. Florida has some of the toughest building codes in the United States, but it would cost too much to make them tougher. We have to keep housing prices low. That's the basis of the state economy."

"Attracting outsiders has always been our primary economic engine," explained journalist and Florida resident Michael Grunwald. The state's prosperity has long rested on "a human pyramid scheme—an economy that relied on a thousand newcomers a day . . . whose livelihoods depended on importing a thousand more newcomers the next day." All those new arrivals need places to live. That means that housing prices, as Gispert said, have to be kept as low as possible, which in turn spurs the building of more and more homes, including in vulnerable coastal areas.

This philosophy was shared at the top of Florida's government: as governor from 1998 to 2006, Republican and presidential brother Jeb Bush was very pro-development. Governor Bush may have felt less alarm about overbuilding along the coast because, like his older brother, he was dubious about the science of global warming. As late as 2009, Bush the younger said in *Esquire* magazine that he was "skeptical" about global warming, largely because of the (supposed) potential of emissions reductions to harm the economy.

The refusal to take climate change seriously instead opened Florida's economy to a different threat. The extra-powerful hurricanes of 2004 and 2005 alarmed the insurance industry, which paid out $250 million in damages for the entire Atlantic coast. Companies responded by dramatically increasing prices and reducing coverage. Many policyholders were dropped altogether; those who could still find coverage had to pay much higher premiums. Some homeowners' rates increased roughly 100 percent over two years, which led many people, especially retirees on fixed incomes—a sizable proportion of Florida's population—to give up their insurance altogether, a terrible risk in a state so susceptible to hurricanes.

By fall 2006 the insurance crisis was the biggest political issue in the state, with staggering economic implications: without insurance, houses can't sell, businesses can't get loans, commerce falters. To entice insurance companies to relax their terms, the state legislature dangled increased subsidies and other incentives. But with memories of the 2004 and 2005 hurricane seasons still fresh and with climate scientists projecting more of the same in the future, insurers wouldn't bite.

So the legislature embarked on a monumentally risky endeavor of its own: it made the state the insurer of last resort in Florida. Henceforth, the state government's Citizens Property Insurance agency would provide insurance to all qualified Floridians, effectively making taxpayers liable for all damages. In the short term, this intervention kept people in their homes and businesses operating. In the medium to long term, it all but promised to bankrupt the state. "If a Category 4 or 5 hurricane hit Tampa, estimates are that it would cause $50 to $65 billion in damages," said Bill Newton, an insurance expert with the nonprofit group Florida Consumer Action Network. "Well, the state's entire annual budget is about $60 billion. So we'd be sunk."

Insurance is essential to modern economic life, so figuring out how to keep it available and affordable in the face of intensifying climate change is a central challenge for adaptation policy. There are no easy answers; Aalt Leusink told me even the Dutch are struggling with this one. But Florida's approach amounts to a huge gamble against the warnings of climate science: the state's economic survival depends on avoiding mega-storms at the very time such storms are projected to become more frequent.

Jim Donelon, the state insurance commissioner of Louisiana, who cautioned his governor against adopting Florida's approach, said Washington must intervene. "The only solution is to get the federal government to do what it did after September 11 and recognize that some risks are too large and costly for the private insurance market to absorb on its own," Donelon told me. When private companies balked at insuring against potential terrorist attacks on San Francisco's Golden Gate Bridge and Chicago's Sears Tower, Congress passed the Terrorism Risk Insurance Act of 2002, which made $100 billion in federal money available as a backstop to private insurance for such structures. Like federal insurance guarantees for nuclear power plants, this $100 billion would be spent only in the event of a catastrophic incident. Donelon advocated that a similar fund be established for cities and other high-value places threatened by climate change, an idea seconded by Tampa's Mayor Iorio.

Such a scheme might succeed in managing the unavoidable risks of the second era of global warming, but what about avoiding the unmanageable? No amount of federal subsidies can make insurance economically feasible for long unless global warming is soon halted and reversed. "If I were the insurance czar of Florida," said Newton, "I'd move on three fronts at once. First, we have got to get serious about cutting greenhouse gas emissions. If global warming isn't stopped, Florida doesn't have a future, period. That said, our risks will go up over the next fifty years no matter what, so we also have to be a lot smarter about what kind of development we allow in coastal areas. Take the Florida Keys. You don't want to shut down the Keys; they're incredibly beautiful and draw tourism from all over the world. But you can say that mobile homes aren't allowed there—they're just too dangerous in a storm. Now, you still need people to work in the tourist hotels and restaurants, so the second thing we have to do is develop alternative low-income housing that is sustainable and resilient. Finally, we need to set up rules for what to do when a community gets wiped out. Which places get rebuilt and which don't? New Orleans is easy. You have to rebuild there; New Orleans is too important to the national economy not to. But in Florida there are lots of places that shouldn't be rebuilt. They're just not valuable enough to the larger society."

Another imperative is reforming the federal flood insurance program. Established in 1968, at the height of President Lyndon Johnson's Great Society, the program "was designed for one purpose: to provide a measure of insurability in communities that were exposed to flood risk," said Tulane's Mark Davis. "It was not intended, though it certainly had the effect, to induce development in risky places." Once federal flood insurance became readily available, in the 1970s, real estate developers and construction firms joined with local politicians to expand settlements well beyond what prudence dictated. As with Florida's Citizens Property Insurance agency, the U.S. government ended up taking on risk that private insurers shunned; meanwhile, it kept premiums artificially low to boot. When floods came, the feds not only reimbursed homeowners, they also offered them fresh insurance, thereby starting the cycle all over again. The message sent was "We will help you build where you shouldn't, we'll rescue you when things go wrong, and then we'll help you rebuild again in the same place," said Paul Farmer, an expert with the American Planning Association.

Likewise, Hurricane Katrina did not deter people in Florida from building and rebuilding as if powerful hurricanes were an anomaly rather than an increasingly likely threat. "When Hurricane Charley came through Charlotte Harbor, it destroyed a whole row of beachfront condos, and now they're rebuilding condos in exactly the same place," said Newton. His voice all but shrieking with incredulity, he added, "You can't do that! That's not getting the message."

Shanghai on the Edge and in Denial

It is not just the American South, with its conservative politics and anti-government mindset, that has not been getting the message. In China, where the Communist Party exercises firm control over most aspects of public life and in theory puts the good of society above individual interests, I found that authorities were doing little so far to anticipate, much less prepare for, the looming impacts of climate change. Perhaps the most vivid—and economically reckless—example is in Shanghai, where a combination of sea level rise and fiercer river and ocean flooding threatens the business capital of China with a disaster that a senior government scientist warned would be as bad as or worse than what New Orleans suffered from Katrina.

Not until the morning I left Shanghai did I fully grasp how vulnerable the city is to climate change. Shanghai is the one Chinese mega-city

I had missed while visiting in 1997 for *Earth Odyssey,* so it was a revelation to see it now. Shanghai was even richer than I had imagined—its streets choked with traffic jams of Mercedeses and BMWs, its shopping districts boasting top-end brands from Europe and the States, its downtown crowded with skyscrapers housing some of the world's biggest companies, all of them intent on riding the magic carpet of endless economic growth that is modern China.

I was keen to take the world's fastest train to the Shanghai airport, even though that required a taxi ride first. The station for the magnetically levitated Maglev train is located in the southeastern suburbs, across the Huangpu River, which runs through the middle of Shanghai. To get there, my taxi ascended one of the steeply pitched entrance ramps typical of Shanghai's freeway system, which rings the city at dizzying heights. I quickly found myself staring into the nearby windows of residential skyscrapers, watching grandmas shuffle across kitchen floors and fathers dress for work while televisions flickered in the next room.

Skyscrapers are everywhere in Shanghai. I tried to obtain an official count, but it was impossible; more were being built all the time. I can report anecdotally that Shanghai's skyline boasts many more skyscrapers than New York's, which, when you think about it, is not surprising. Shanghai's population is 19 million people, more than twice as large as New York's. The only way that many people can live in so small an area is to colonize the sky and stack people on top of one another.

Skyscrapers are not the best places to be in the second era of global warming. They are utterly dependent on electricity, above all to run the elevators, and power outages may well become more common in the future as temperatures rise and storms grow stronger and more frequent. As noted earlier, hotter weather causes power blackouts for two reasons: the heat stresses generators and transmission lines, and power demand spikes as customers turn on fans or air conditioners to cope. If a blackout lasts more than a day or two, skyscrapers will become forbidding places. How do you get to your apartment without elevators? If you live on the lower floors, you might take the stairs. But if you live more than eight or ten floors up—and most of Shanghai's residents clearly belong in that category—could you really climb that many stairs? How many times a day? What about your elderly neighbor?

The train ride to the airport was thrilling, I must say. The train accelerated smoothly and very soon was traveling very fast. We flashed across an expanse of flat, swampy land bisected by narrow canals, and seven minutes later we were at the airport. I wobbled off the train, still feeling the rush of traveling 269 miles an hour. I took an escalator upstairs and found myself facing a gigantic wall of glass that stretched the length of the terminal. The view beyond was of the airport runways and, just beyond them, the South China Sea, where a cargo ship steamed slowly past.

My jaw dropped: the runways were at essentially the same elevation as the sea. It reminded me of the airports back in San Francisco and Oakland, only worse. It was bad enough that Shanghai's airport occupies low-lying ground that is as flat as a pool table (which made sense: Shanghai sits in the floodplain of one of the world's four mightiest rivers, the Yangtze). But Shanghai's airport is also perched at the edge of a sea that is notorious for its typhoons.

Sea level rise is clearly going to pose enormous problems for Shanghai in the not-too-distant future, but if I lived there I'd worry more about the short-term risks of big typhoons. According to data compiled by the IPCC, three feet of sea level rise would put an estimated twenty-eight square miles of Shanghai underwater. The storm surge from an average typhoon could easily be three times higher than that, which presumably would inundate a much larger area. (I could not locate rigorous estimates for such a scenario—it may be that such calculations have not been made.) "We average two to three typhoons in this area a year," said Mao Weide, the former director of the Shanghai Water Authority, which handles flood protection, water supply, and water treatment. The city also experiences "strong winds, high tides, and torrential rain two years out of three, and some of these have been quite severe," he added. In addition, there is the threat of flooding from the Yangtze, no small matter.

To protect the city, an extensive system of levees had been built, some dating back to ancient days but most constructed after the Communists came to power in China in 1949. "Since Liberation the government has given great importance to flood control," said Mao, whose eminence was underscored by the fact that the five other experts attending our interview said hardly a word the entire two hours. "The government built five hundred kilometers of dikes along the Yangtze and Huangpu rivers and the shore of the South China Sea." Mao said that the suburban areas of Shanghai, where the airport is located, had been given 1-in-100-years protection against floods. Most urban areas were given 1-in-200-years. But the urban area along the Huangpu River, the heart of Shanghai, boasted 1-in-1,000-years protection, according to Mao.

All this seemed plausible enough until I went to see the dikes for myself. Before leaving for China, I had read a report by the Organization for Economic Cooperation and Development that ranked Shanghai's flood defenses among the best in the world—comparable, the OECD said, to London's. But whoever wrote that report either did not look very closely at the defenses or simply took the government's word for their soundness. The reality I discovered on the ground—or perhaps I should say along the waterline—was much less reassuring.

Mao and his colleagues had advised that the inner core of Shanghai be protected with dikes that were nineteen feet tall to provide 1-in-1,000-years protection. Some of the dikes I inspected did meet that standard. I began my investigation on the western side of the Huangpu River—home to the Bund, a stretch of riverfront boasting one grand building after another, a legacy of the old imperial era when Western merchants and governments competed to construct the most ostentatious structures possible. Reaching the riverfront was not easy; access was restricted by construction projects preparing Shanghai to host the 2010 World Expo. When I finally found an opening through a smelly port house, I emerged at the water's edge to find the river encased by a sturdy-looking seawall made of stone and concrete and measuring about three feet thick. I checked again farther north, along the promenade that curves beneath the Bund, and checked again yet farther north, near Suzhou Creek, where the old British and French consulates were located. I found the same at all three spots: the thick stone seawall was about thirteen feet above the waterline. True, this was less than the nineteen feet the authorities had cited, but my visit came at high tide, which could account for much of the discrepancy.

But across the river, in the trendy business district of Pudong, I discovered another story entirely. Pudong is the fastest-growing part of Shanghai, the place where most multinational corporations are located and many of the expatriate community live. The riverfront was much easier to access on the Pudong side; a short walk from the Lujiazui sub-way station brought me to a tourist area with a small sloping lawn, a restaurant, and a walkway along the water. It was a sunny afternoon, and the area was crowded with families and young lovers, nearly all of them Chinese. I stepped to the water's edge, looked down, and was shocked. The murky green water was no more than four feet below the walkway; a medium-sized storm surge would easily put this area under-water. "This is 1-in-1,000-years protection?" I muttered.

From a historical perspective, it makes sense that Pudong is less pro-tected than the Bund side of the river. In olden days, all the money and power in Shanghai resided on the Bund side; Pudong was sparsely set-tled until about twenty years ago. But from an economic perspective, this was preposterous: behind these puny dikes are many of the brains and businesses that animate the Chinese economic miracle, not to men-tion millions of ordinary residents.

Neither the authorities I interviewed nor the Shanghai natives and expats I met seemed worried, or even informed, about any of this. These flood defenses will be increasingly stressed as global warming intensi-fies in the years ahead. But Mr. Mao was confident almost to the point of arrogance that everything remained under control, and if his five col-leagues disagreed with him, they were unwilling to say so. But then Mao seemed unconvinced that global warming was real in the first place. He did agree that the earth is warming, but he was not sure humans had much to do with it. When I asked whether the Shanghai Water Author-ity incorporated projections of climate change into its calculations of future storm strength and sea level rise, he explained, "We can do our projections only on the basis of historical data and monitoring of cur-rent events." In other words, no.

In retrospect, I shouldn't have been surprised. Back in Beijing, a top government scientist had told me that neither Shanghai authorities nor central government leaders appreciated just how much danger Shang-hai is in. This scientist was extremely uncomfortable criticizing cen-tral government leaders—they had long read his reports—so I will not identify him by name. But I can say that he had done enough initial research into what climate change implies for sea level rise and storm strength that he had become alarmed. He was most concerned by the threat to China's coasts and Shanghai, the most economically wealthy, dynamic places in China. Specifically, the scientist feared that Shanghai could suffer a Katrina-like disaster.

"Katrina was a terrible, terrible thing," he said, grimacing. He had been traveling abroad when Katrina struck and thus had had the benefit of watching television news unfiltered by Chinese government censors. "Personally, I felt Katrina was even worse than September 11," he added. "That destroyed two buildings, but Katrina destroyed an entire city. China must take all measures necessary to avoid such an event."

The scientist said he had tried to sound an alarm by appealing to top figures of the central government. He presented his concerns, which were shared by a number of his colleagues, and urged that the government fund a study to determine the actual risks and to identify adaptation strategies for Shanghai. But he found little interest. "Maybe they think the [existing] dikes are enough," he said of the central government leaders. "But we think they are not enough."

"But Shanghai is very, very important to China," I said. "Can't you make the top leaders understand?"

"We try, we try," he said, shaking his head in frustration. But he and his colleagues were caught in a catch-22: they lacked definitive scientific proof of what climate change augured for Shanghai, and without this proof they could not convince the leadership to fund the additional studies needed to produce such proof. "So for now," he concluded sadly, "we must simply hope that Shanghai does not have an experience like New Orleans did."

"I'd Rather Be Smart Than Lucky"

The next fairy tale Chiara fell in love with after *The Nutcracker* was *Peter Pan*. Written for the stage in 1904 by the Scottish author J. M. Barrie, *Peter Pan* is set partly in Victorian London but mainly in Neverland, an imaginary island reached by flying toward "the second star to the right and straight on 'til morning." As with *The Nutcracker*, once Chiara grasped the essence of the story, she took great joy in acting it out over and over again. She usually played Peter, while I took the role of his archenemy, Captain Hook, complete with a snarling voice that sometimes frightened Chiara a little. We had pretend sword and cannonball fights, and I would flee in terror whenever the pretend crocodile drew near. Chiara loved all this so much that for her third Halloween, she dressed up as Peter Pan, in a green felt shirt and leggings, matching cap, and brown felt dagger. I was Hook, with a military greatcoat, a flamboyant hat, and a fake hook where my left hand should have been.

Chiara didn't seem to pay much attention to it—she was only three, after all—but the fact that Peter repeatedly refused to grow up and join the adult world struck a chord with me. Working on this book, I was forever coming across upsetting projections about how climate change would endanger my daughter's future, whether through the cyclones and sea level rise discussed in this chapter, the extinction of polar bears and other plant and animal species, or—this one landed in my in box this morning—the rapid thawing of tundra in Canada and Russia that makes runaway global warming more likely. In the face of all that, the fantasy that Chiara, like Peter Pan, might choose *not* to grow up held a certain fascination for me. Though it's true, as the cliché puts it, that every age your child goes through is a good age, I confess that selfishly I liked the idea that Chiara might stay little for longer and thus delay the sad day when she would leave home to live on her own.

But Chiara, as is her way, set me straight on all this. We had a little game we sometimes played about how big she was getting. Whenever she did something that demonstrated a new level of physical or emotional maturity—clearing her plate from the table, say, or topping three feet tall at the doctor's office—I would first tell her how proud of her I was. Then I'd put on a pretend-sad face and say, "But I'm afraid you're getting too big." Knowing what was coming next, Chiara would start to grin. "But I know how to fix that," I continued. "You're three years old now. How about on your next birthday you turn two? You're just growing up too fast." She would look at me, big smile now lighting her face, and shake her head side to side. "No, Daddy," she replied. "Kids can't grow down. They can only grow up."

Kidding aside, I look forward to seeing Chiara grow up; I can't wait to meet the person she will be as a young adult. Yet I can't forget the terrible difficulties that lie in wait for her and countless other children of current and future generations, thanks to humanity's failure to reverse global warming sooner. So the only responsible course of action at this point is to acknowledge the path we're on, take a deep breath, and begin pressing our governments, our businesses, our civic institutions, and ourselves to do all we can to avoid the unmanageable and manage the unavoidable.

I'll address avoiding the unmanageable in the last chapter of this book, but as far as managing the unavoidable, the first step is to jettison the assumption that luck will see us through. At the moment, an enormous number of people and institutions around the world are making exactly this assumption, often without knowing they are doing so. Indeed, all those who are living in harm's way but are not actively pursuing adaptation are, in effect, counting on luck to keep them safe. To be sure, such a strategy sometimes works in the short run. In New Orleans, for example, efforts to rebuild the city's flood defenses after Hurricane Katrina stumbled badly, for reasons described earlier in this chapter. It looked as if the city might again pay a terrible price when, in late August of 2008, exactly three years after Katrina, Hurricane Gustav appeared off the coast of Haiti. After blasting that impoverished island nation, Gustav weakened, only to regain strength a few hours later. As it headed west into the Gulf, Gustav registered as a Category 4 storm. Authorities in Louisiana and other Gulf states ordered residents to leave, unleashing one of the largest evacuations in history, and one that by and large went well—a sign that at least some lessons were learned after Katrina. But the key element was luck. Gustav weakened and shifted course as it crossed the Gulf. By the time it made landfall in the less populated western side of Louisiana, it was a Category 2 storm. Nevertheless, it still killed 48 people in the state and 112 more in all affected regions.

"We [in New Orleans] got lucky this time," Davis of Tulane later told *Time,* explaining the relatively low death toll. "I like being lucky," he added. "But at some point we have to get smart."

So what does it mean to be smart in the second era of global warming? The cases examined in this book so far—New Orleans, the Netherlands, Tampa Bay, Shanghai, King County, London, and the rest—suggest a number of conclusions.

First, being smart about climate change requires learning about it: understanding the basic science and its implications, despite one's ideological dispositions or economic interests. It is particularly vital to ascertain the local and regional climate impacts to which one must adapt. King County, Great Britain, and the Netherlands are ahead of the rest of the world on adaptation partly because they embraced the science of climate change many years ago and were prudent enough to respond accordingly. New Orleans, Florida, Shanghai, and many other places remain more vulnerable than necessary because their political and economic leaders, many citizens, and even some scientists continue to doubt that burning fossil fuels poses grave risks to their future.

Being smart also means planning ahead. The human and economic disaster that followed Hurricane Katrina illustrated what can happen when government fails to plan ahead, just as the improved flood defenses and climate-friendly urban planning found in King County and the Netherlands show what can be accomplished when governments ask and answer the climate question. Individuals have responsibilities, too; among other things, they must be prepared to evacuate from high-risk locations and not dawdle when authorities order them to go. But the central role inevitably belongs to government. Only a government can oversee and operate evacuation plans. Only a government can organize, fund, build, and maintain levees and seawalls to keep unwanted water out and dams and pipelines to bring fresh water in. Only a government can implement the myriad decisions about public health, land use, species protection, and other issues that climate change will force all societies to confront.

Governments should not make these decisions alone, however. The most enduring, effective decisions tend to be made when the general public, scientists and other recognized experts, affected businesses and institutions, and other relevant stakeholders are involved in the process: not only is more expertise brought to bear, but the stakeholders are more likely to support the ultimate decision. As Aalt Leusink said of Dutch efforts to secure the widest possible participation in adaptation planning, "We really want people's input. In fact, we cannot do this without them." The dictatorial powers that government wields in a one-party state may seem attractive—in Shanghai, for example, I learned that the authorities had recently ordered countless highly polluting motorbikes off city streets virtually overnight. But what happens when an authoritarian government makes the wrong decision? Checks and balances—transparent decision-making processes, news media free to keep the public informed and the government honest, a citizenry unafraid to criticize and demand better from its leaders—minimize government errors and produce superior decisions. Governments often need to be pushed to do the smart thing. Citizens must be free to do the pushing without getting thrown in jail or worse. Which is why human rights, freedom of the press, and other elements of an open society are critical to coping with climate change.

Governments must also be well funded, which means that taxes must be paid. Governments must be held to high standards, with zero tolerance for wasteful spending or corruption, but citizens and businesses must be mature enough to recognize that material things of value cost money. If we want our governments to help protect our communities from climate change, they must have the funds needed to do the job. Government failed disastrously in New Orleans largely because the relevant agencies were underfunded and demoralized after decades in which the guiding assumption in Washington was that government is evil, taxes are theft, and regulation is unnecessary. Raising taxes is assumed to be suicidal for politicians, but that has not been the case in King County or the Netherlands, where leaders not only raised taxes but got reelected afterward. The secret? Honesty and follow-through. Both Ron Sims and the Dutch government told their electorates why the additional taxes were necessary and then used the revenues to do what they promised. In other words, they treated citizens like grownups.

All of these lessons underscore the importance of the social context of adaptation—the mix of public attitudes, cultural habits, political tendencies, economic interests, and civic procedures that shape a given society's way of addressing public issues. Social context is often deeply rooted and therefore difficult to change, especially in the short period of time we have to prepare against climate change. It's all very well to urge a community or country to reform its social context—to accept the science of global warming, embrace advance planning and public participation, and regard taxes and government as necessary tools rather than despised oppressors. But climate change will not wait for such reforms. The upshot is that places that already have a favorable social context for adaptation will likely do better at living through the storm. Places that shun such considerations are more likely to lose. Of course, such places could get lucky. But even in fairy tales, luck goes only so far.

7 | *In Vino Veritas:* The Business of Climate Adaptation

> When I give talks to companies, I tell them that climate change is like the Internet. It suddenly appears one day, it changes everything, it gets bigger every year, and you have to learn how to make money from it or you're going to get eaten for lunch.
>
> —PAUL DICKINSON, CEO, Carbon Disclosure Project

WAYNE LEONARD STARTED worrying about climate change years before Hurricanes Katrina and Rita cost his company $1.5 billion. Leonard is the chief executive officer of Entergy, which provides electricity to 2.7 million customers in Louisiana, Texas, Arkansas, and Missouri. In 2001, under Leonard's leadership, Entergy became the first electric utility in the United States to pledge to limit its greenhouse gas emissions. Cynics scoffed that it was an easy gesture to make: much of Entergy's electricity came from nuclear power plants, which produce no greenhouse gases. But Jeff Williams, the company's director of climate consulting, maintained that his boss had other motivations. As CEO of the only Fortune 500 company headquartered in New Orleans, said Williams, Leonard understood how vulnerable his company and the rest of Louisiana were to sea level rise and hurricanes, and he hoped Entergy's example would encourage other companies and the U.S. government to get serious about fighting global warming.

Hurricane Katrina turned Leonard's fears into reality. Flooding from Katrina put two of Entergy's power plants out of commission, along with transmission lines and a natural gas distribution system. "More than that, [Katrina] risked the economic viability of the entire metropolitan area [of New Orleans]. More than half of the pre-Katrina population hasn't returned home yet," said Williams in 2007. "Those are our customers." The economic costs of Katrina, not to mention the human ones, showed why reducing greenhouse gas emissions was not only an environmental but also a business necessity. He added: "Katrina cost the U.S. economy $200 billion. Yet every time you hear about climate legislation in the news, it's said that cutting emissions will kill the economy. Not once do I hear what the costs of adapting to [unmitigated] climate change would be. Katrina shows those costs would be huge."

In the years following Hurricane Katrina, Wayne Leonard continued pressing for emissions reductions, but he also did something even more unusual for a U.S. business leader: he embraced adaptation. In October 2009, two months before the Copenhagen climate summit, he spoke at the Obama White House as part of a business delegation urging passage of the Waxman-Markey climate bill. Warning that other coastal cities could face disasters worse than the one New Orleans suffered if climate change is not stopped, Leonard made the investors' case for taking action. "We condemn Wall Street for taking risks with our money," he said, ". . . but at the same time we're taking exactly the same kind of risks, with no upside whatsoever, with regard to our climate, failing to practice even the most basic risk management techniques." Risk management, in Leonard's view, includes both mitigation and adaptation. Investing in emissions reduction "pays back many times over," the CEO argued, notably in "reduced adaptation costs in the future" and in boosting general economic prosperity "through transferring technology around the world and creating jobs at home." In Entergy's case, adapting to climate change meant spending millions of dollars to relocate a data center and a transmission facility, moving them well inland. The goal, Williams told me, was to reduce the risk and costs of any business disruptions caused by another hurricane. "Hurricanes are part of life on the Gulf coast," Williams said, "but they don't have to cost you $1.5 billion every time they happen."

As important as strong government is, no society will avoid the unmanageable and manage the unavoidable of climate change if its business community does not get involved. "The majority of adaptation (or maladaptation) will come from everyday decisions in the private sector," Maya Forstater, Saleemul Huq, and Simon Zadek point out in *The Business of Adaptation,* a paper published by the International Institute for Environment and Development in London. Corporations and individual entrepreneurs can make the right kind of investments to build a climate-friendly and climate-resilient economy or the wrong kind. Governments set the rules of the road in terms of economic incentives and regulations; they also oversee large-scale infrastructure projects such as building seawalls and upgrading health systems. But the private sector controls the vast majority of a society's wealth, sets its overall economic direction, and provides goods and services vital to everyday life. If the companies that grow our food, deliver our water, produce electricity, construct roads and buildings, supply medicines, manufacture clothes, and sign our paychecks cannot continue those functions, society's very existence is in peril.

Statements like these will be seen as truisms before long, but few recognize them as such today. For most business leaders, coping with climate change means reducing emissions (or at least giving lip service to the idea) and nothing more. "[Mitigation] is a legitimate topic of conversation for business leaders now. In fact, you start to feel a bit embarrassed if you don't have someone in charge of it," says Steve Howard, the CEO of the Climate Group, an NGO that works with corporations worldwide to advance climate protection. But the vast majority of businesses have no knowledge of or involvement with adaptation.

It's odd, because adaptation is much more directly rewarding than mitigation is. Mitigation can make a firm money and burnish its reputation, but the real benefits accrue to civilization as a whole, in the future: fewer emissions mean the planet will experience less climate change beginning approximately thirty years hence, because of the climate system's inertia. What's more, civilization benefits only if others cut their emissions as well, making mitigation partly an act of faith (unless governments enforce mandatory reductions). Adaptation, by contrast, serves the short-term interest of the company doing the adapting, and it does so even if others do not adapt. Nevertheless, it seems not to have occurred to most companies that climate change poses real threats to their future. "The physical risks of climate change are often over-looked by business," observe Frances G. Sussman and J. Randall Freed in *Adapting to Climate Change: A Business Approach,* a survey of corporate practices prepared for the Pew Center on Climate Change in Washington. Those risks include "business interruptions, increased investment or insurance costs, or declining . . . value, return, and growth."

"Most of us are not very good at recognizing our risks until we are hit by them, and people who run companies are no different," explained Chris West of the UK Climate Impacts Programme. Even bankers, who are supposed to understand risk better than most, often miss the point. "One banker told me, 'We do mortgages; why should we worry about climate change?'" West recalled. "I asked what they mortgaged. He said, 'Properties.' Well, properties are sure to be affected."

Under West's leadership, UKCIP has done as much as any organization in the world to awaken businesses to the new realities of climate change. At first glance, working with private companies seems a leap for West, who earned his doctorate at Oxford in zoology and spent the next twenty-odd years working as a British government scientist, specializing in the protection of endangered species. But at UKCIP, the species West is trying to save is his own, and the insights of a zoologist turn out to be quite useful. Adapting to changing circumstances is, after all, the essence of evolution—and of success in the modern economic market-place. West is fond of quoting Darwin: "It is not the strongest of the species that survives . . . nor the most intelligent that survives. It is the one that is the most adaptable to change."

A Global Warming Poster Child

Every business on earth will feel the effects of climate change in the years ahead, but few will feel it as soon and as acutely as the wine industry. Because wine grapes are extraordinarily sensitive to temperature, the wine business amounts to an early warning system for the problems that all food crops—and all businesses—will confront as global warming intensifies. *In vino veritas,* the Romans said: "In wine there is truth."

The truth nowadays is that the earth's climate is changing much faster than the wine industry, and virtually every other business on earth, is preparing for.

All crops need favorable climates, but few are as vulnerable to temperature and other extremes as wine grapes; a few degrees of difference in temperature, especially at critical times of the growing season, can make the difference between a superb wine and an undrinkable one. And the economic consequences can be staggering. "There is a fifteen-fold difference in the price of Cabernet Sauvignon grapes that are grown in Napa Valley and Cabernet Sauvignon grapes grown in Fresno," in California's hot Central Valley, said Kim Cahill, a scientist studying climate change impacts as a postdoctoral fellow at the University of California at Davis, America's premier research institution for all things wine-related. "Cab grapes grown in Napa sold [in 2006] for $4,100 a ton. In Fresno the price was $260 a ton. The difference in average temperature between Napa and Fresno was 5°F. Obviously, there are differences in soil and management as well, but the climate has a major impact on price."

Numbers like that help explain why climate change is poised to clobber the global wine industry, a $50-billion-a-year enterprise whose decline would also cause collateral damage to the much larger industries of food, restaurants, and tourism. In France, for example, the rise in temperatures may render the Champagne region too hot to produce fine Champagne. The same is true for the legendary reds of Châteauneuf-du-Pape, where the stony white soil's ability to retain heat, once considered a virtue, may now become a curse. The world's other major wine-producing regions—California, Italy, Spain, South Africa, Australia—are also at grave risk. If current trends in greenhouse gas emissions and temperature rise continue into the future, the "premium wine grape production area [in the United States] . . . could decline by up to 81 percent by the late 21st century," a team of scientists including Gregory Jones, a professor and research climatologist at Oregon Southern University, wrote in a study published in the *Proceedings of the National Academy of Sciences* in 2006. The culprit was not so much the rise in *average* temperatures but an increased frequency of extremely hot days, defined as above 35°C (95°F). If no adaptation measures were taken, these increased heat spikes would "eliminate wine grape production in many areas of the United States," the scientists wrote.

In theory, winemakers can defuse the threat by simply shifting production to more congenial locations. Indeed, champagne grapes have already been planted in England and Denmark and some respectable vintages harvested. But there are limits to this as a strategy for the industry. After all, temperature is not the sole determinant of a wine's taste. What the French call *terroir*—a term that refers above all to the soil and physical environment of a given region but also includes, in some definitions, the cultural knowledge of the people who grow and process grapes there—is crucial. "Wine is tied to place more than any other form of agriculture, in the sense that the names of the place are on the bottle," said David Graves, the cofounder of the Saintsbury wine company in southern Napa Valley. "If traditional sugar beet growing regions in eastern Colorado had to move north, nobody would care. But if wine grapes can't grow in the Napa Valley anymore, which is an extreme statement but let's say so for the sake of argument, suddenly you have a global warming poster child right up there with drowning polar bears."

A handful of climate-savvy winemakers such as Graves are trying to rouse their colleagues to action before it is too late, but to little avail. Many winemakers are actually rejoicing in the higher temperatures of recent years, because they believe the extra heat has helped produce some of the best vintages in memory. "Some of the most expensive wines in Spain come from the Rioja Alta and Rioja Alavesa regions," Pancho Campo, the director of the Wine Academy of Spain, told me. "They are getting almost perfect ripeness every year now for Tempranillo. This makes the winemakers say, 'Who cares about climate change? We are getting perfect vintages.' It is very difficult to tell someone who has accomplished something in his vineyard, 'This is only going to be the case for another few years.'" I heard much the same from producers I interviewed in Italy and in California's Napa and Sonoma valleys. If this is climate change, winemakers seemed to be saying, bring it on.

"The Character of Our Wines Was Changing"

Perhaps the leading exception to the trend is Alois Lageder, whose family has made wine in Alto Adige, the northernmost province in Italy, since 1855. The setting, at the foot of the Alps, is majestic. Looming over the vines are massive outcroppings of black, white, and gray rock interspersed with flower-strewn meadows and wooded hills that inevitably call to mind *The Sound of Music*. Inspired by the legendary California winemaker Robert Mondavi, Lageder led Alto Adige's evolution from its jug wine past to producing some of the best white wines in Italy today. (Alto Adige produces only 0.7 percent of Italy's wine but is responsible for 10 percent of its premium wine production.) In October 2005, Lageder hosted the world's first conference on the future of wine under climate change. His personal prestige helped draw a good-sized crowd; some four hundred growers, winemakers, and other experts from Alto Adige and neighboring provinces attended. "We must recognize that climate change is not a problem of the future," Lageder told his colleagues. "It is here today and we must adapt now. Winemakers must be willing to experiment and adapt if they wish to survive under climate change."

Embracing experimentation himself, Lageder has increased his wine's quality even as he introduced biodynamic farming. Grounded in an almost mystical belief in the superiority of natural processes, biodynamic farming goes a big step beyond organic farming. Compost replaces fertilizer; ground cover and mulch substitute for irrigation; in some cases, planting and other key activities are timed to the cycles of the moon. Skeptics dismiss it as woolly-headed nonsense, but "this was how all of our ancestors practiced agriculture, and it has many advantages," said Lageder, adding, "My mother always used biodynamic methods in her own garden."

To illustrate the point, Lageder, a slender man with receding brown hair, took me for a tramp in his vineyards. "Look here," he said, kneeling down and digging his index finger into the earth beneath vines that would produce his coveted Löwengang label Cabernet Sauvignon. "See how easy it is for me to penetrate this soil? This is an indication of how healthy [the soil] is with lots of humus; the water, air, worms, and microorganisms pass through it and let it breathe. Now, try to do the same over here," he said, pointing to vines fifteen feet away that belonged to a neighbor who farmed conventionally.

I knelt and pressed my finger into the neighbor's soil, but it was unmovable—dry and crusty. Lageder then pinched off a leaf from his neighbor's vines and a leaf from his own and placed them on the dashboard of his car. It was a warm day, and when we returned to the car an hour later, his neighbor's leaf was wilted, but Lageder's still had plenty of life in it. "Biodynamic farming is helpful in storing water in the soil, and this keeps the vines healthier," Lageder explained. "We have found we have not had to irrigate during hot summers when our neighbors did have to irrigate. So maybe this is a partial solution to the problems with water scarcity that climate change will bring. Vines will have to be more resilient if they are to adapt to climate change."

As it happens, Alto Adige is the location of one of the most dramatic expressions of modern global warming: the discovery of the so-called Iceman—the frozen remains of a hunter who had lived in the Alto Adige region some 5,300 years ago. The Iceman's corpse was found in September 1991 in a mountain crevasse just meters from the Austrian border. His body was almost perfectly preserved—even the skin was intact—because it had lain beneath mounds of snow and ice ever since shortly after his death (a murder, forensic investigators later concluded from studying the trajectory of an arrowhead lodged behind his left shoulder). The Iceman, now housed in a museum in Bolzano, provided the most detailed and lifelike example ever discovered of how our ancient ancestors looked and lived (the mummies of Egypt are nowhere near as well preserved), and the discovery soon became an international media phenomenon. But the Iceman never would have been found without global warming, said Hans Glauber, the director of the Ecological Institute of Alto Adige. "Temperatures have been rising in the Alps about twice as fast as in the rest of the world," Glauber explained. "These higher temperatures have been melting the snowpack, and one result was the emergence of the Iceman."

Lageder had already heard about global warming and felt compelled to take action. "When I was a kid, the harvest was always continuing after November 1, which was a cardinal date," he told me. "Nowadays, we start between the fifth and tenth of September and finish in October." His first step was to rebuild his winery according to ecological principles. It wasn't easy—"I had incredible fights with my architect about wanting good insulation," he said—but by 1996 his winery relied solely on solar and geothermal energy, making it the first privately financed solar installation in Italy. Care was taken to integrate these ultramodern technologies into the existing character of the site; during a tour of the facility, Lageder and I emerged from a dark fermentation cellar with its own wind turbine into the bright sunlight of a gorgeous fifteenth-century courtyard. Going green did make the renovation cost 30 percent more, Lageder acknowledged, "but that just means there is a slightly longer amortization period. In fact, we made up the cost difference through increased revenue, because when people heard about what we were doing, they came to see it and they ended up buying our wines."

The record heat of the summer of 2003 sparked new alarm on Lageder's part about global warming. Excess heat raises the sugar level of grapes to potentially ruinous levels. Too much sugar can result in wine that is too heavy and alcoholic—wine known as "hot" or "jammy." Higher temperatures also increase the risk of diseases and pests, because fewer of the latter than normal die off during the winter chill. White wine grapes, whose skins are less tolerant of heat, face particular difficulties as global warming intensifies. "In 2003, we ended up with wines that had between 14 and 16 percent alcohol," Lageder recalled, "whereas normally they are between 12 and 14 percent. The character of our wines was changing."

A 2 percent increase in alcohol may sound like a tiny difference, but the effect on a wine's character and potency is considerable. "In California your style of wine is bigger, with alcohol levels of 14 and 15, even 16 percent," Lageder continued. "I like some of those wines a lot. But the alcohol level is so high that you have one glass and then"—and here he slashed his hand across his throat—"you're done; any more and you will be drunk. In Europe, we prefer to drink wine throughout the evening, so we favor wines with less alcohol. Very hot weather makes that harder to achieve."

There are tricks grape growers and winemakers can use to lower alcohol levels. The leaves surrounding the grapes can be allowed to grow bushier, providing more shade and reducing how sugary the grapes become. Vines can be replaced with different clones or rootstocks that are more resistant to heat. During processing, a centrifuge can separate the alcohol from the fermented juice and return only some of the alcohol to the juice before bottling. Growing grapes at higher elevations, where the air is cooler, is another option. So is changing the type of grapes being grown. (Among reds, Pinot Noir has a relatively low tolerance for heat, Cabernet Sauvignon a high one. White wine grapes, on the other hand, have very limited adaptability to higher temperatures.)

But it takes time and money to make these changes, and most of the wine industry is not yet interested in making the investment. Two years after his wine and climate change conference, Lageder commented, "Some of my colleagues may admire my views on this subject, but few have done much. People are trying to push the problem away, saying, 'Let's do our job today and wait and see in the future if climate change becomes a real problem.' But by then it will be too late to save ourselves."

"I would estimate that 15 to 30 percent of the [global] industry are informed and concerned about climate change, but only 5 percent are speaking out about the problem and trying to address it," said Greg Jones, the Oregon Southern University professor who had visited most of the world's wine regions in recent years, often more than once, and had spoken with scores of producers.

"There's a lot of whistling past the graveyard," said Graves of Saintsbury, one of the very few figures in the industry conversant with the science of global warming. "I don't think people understand the magnitude of change that is built into the system, the fact that temperatures are going to keep going up for a long time no matter what we do."

Jim Verhey, a trim, white-haired gentleman who is a director of the Napa Valley Grapegrowers Association, told me in 2008 that "global warming may not have any truly significant impact on Napa Valley grape quality for hundreds of years." After all, he said, temperatures already can vary as much as 15°F between the northern and southern ends of Napa Valley today, yet both areas produce spectacular wines. Napa, he added, is blessed with ideal soils for wine, and those soils are not going away just because of climate change. Napa also benefits, Verhey said, from the cooling effect of nearby San Francisco Bay. Finally, Napa's growers and winemakers also have the money and skills needed to counter the effects of excessive heat. Pointing to a nearby row of vines, he said, "If temperatures go up, we could change the vines' canopies so the grapes get more dappled light rather than direct sun." Verhey conceded that winemakers were "struggling more now to harvest the best-quality grapes we can because of the 100°F-plus heat spikes now occurring more often. Extreme weather, rather than climate change, has become a significantly greater factor, significantly greater. But I'm optimistic. We can do this."

When I checked in with Verhey a year later, though, he was less chipper. Though still claiming to be optimistic that the wine industry could handle climate change, he admitted he was "frustrated" that the necessary effort wasn't being made. "We talk a lot about [climate change], but I wonder how committed we as an industry are to addressing it," he told me. Noting that he had just spent $100,000 to install a new trellis system to give his vineyards more shade, he said, "There are a handful of people who are pushing the envelope as hard as they can. But I don't see a lot of others taking these measures yet."

That included a colleague Verhey had praised as one of "the smartest young guys in the business," Jon Ruel of Trefethen Family Vineyards. Ruel too was confident of the future of Napa Valley wine, saying, "If anybody can handle climate change, it's us." But he did not seem to regard climate change as much of a threat. As he prepared to replant part of the 550 acres he cultivated in the heart of Napa Valley, Ruel told me, "We're moving away from Cabernet Sauvignon to Merlot and Malbec, which perform better on our ranch. That's interesting, because if I were convinced that our climate was going to be getting hot, I wouldn't do that. I guess I just don't believe we'll get that much hotter here, at least not in the next twenty years."

"Inevitably, Some Existing Businesses Will Fail"

If the wine industry does not adapt to climate change, life will go on — with less conviviality and pleasure perhaps, but it will go on. Fine wine will still be produced, most likely by early adapters such as Lageder, but there will be less of it. By the law of supply and demand, that suggests the best wines of tomorrow will cost even more than the ridiculous amounts they often fetch today. White wine is particularly threatened and may well disappear from some regions. Climate-sensitive reds such as Pinot Noir are also in trouble. It's not too late for winemakers to save themselves through adaptation, but to see an industry with so much incentive to act choosing to dawdle is disconcerting. If the wine industry isn't motivated to adapt to climate change, what businesses will be?

The answer seems to be, very few. Even in Britain, where the government is vigorously championing adaptation, the private sector lags well behind in understanding the adaptation imperative, much less implementing it. "I bet if I rang up a hundred small businesses in the UK and mentioned adaptation, ninety of them wouldn't know what I was talking about," said Gareth Williams, who worked for the UKCIP and later for the NGO Business in the Community, encouraging and assisting businesses in the northeast of England to prepare for the storms and other extreme weather events that scientists project for the region. "When I started this job, I gave a presentation to heads of businesses," said Williams, who had spent most of his career in the private sector before joining UKCIP and knew how to speak with business people. "I presented the case for adaptation, and in the question-and-answer period, one executive said, 'We're doing quite a lot on adaptation already.' I said, 'Oh, what's that?' He said, 'We're recycling, and we're looking at improving our energy efficiency.' I thought to myself, 'Oh, my, he really didn't get it at all. This is going to be a struggle.'

"Most businesses aren't prepared for *current* extreme weather, much less what climate change is going to throw at them," Williams added. "They don't have business continuity plans for keeping their insurance in place, for making sure their records are backed up off-site regularly. What would the implications be of cutoffs of water and electricity supplies? Most businesses would go completely to pieces."

Nevertheless, the British government, through UKCIP, is continuing to work the problem, trying to help business leaders realize that adaptation is in their own self-interest. "When we talk to businesses, we try to put a different slant on the discussion of sustainability," said West. "True sustainability is not just about being nice to people and nature. It's about dealing with environmental threats that can put you out of business."

UKCIP's outreach to the business community began in 2004, when it collaborated with the Association of British Insurers to survey attitudes about climate change among some of the nation's leading trade associations and professional bodies. A year later, UKCIP published *A Changing Climate for Business*, a report that explained in clear, urgent, but not alarmist language how climate change threatens individual businesses and the economy as a whole and what companies can do to reduce their vulnerabilities. The report, updated in 2009 and available at the UKCIP website (http://www.ukcip.org.uk), is the single best resource I have come across for businesses trying to understand and pursue adaptation.

"When we talk with businesses about climate change, we don't start with, 'Here's a big new problem you have to deal with,'" West explained. "We start with where they are: 'Where are your business operations vulnerable? Have you thought about how the weather can affect that?'" To counter the tendency—widespread among executives—to dismiss climate change as a distant future danger, UKCIP urges businesses to "consider their exposure and vulnerability to current weather events as a stepping-stone to consideration of future, potentially more severe, impacts as the climate changes." To dispel gloominess, UKCIP makes a point of highlighting both the threats and the opportunities climate change presents, though scientists routinely emphasize that the former will greatly outnumber the latter. No matter, says West; the goal is to get the conversation started.

Once again, social context is vital; it has an especially powerful impact on a business's capacity to embrace and practice adaptation. "It's really a question of how well an organization copes with a new idea," West added, "because adaptation is still a new idea. Organizations that do well with new ideas tend not to be hierarchical and are often relatively young, so things are not set in stone—the organization can value information that may come from a so-called junior person and get the information where it needs to go. Whereas older, more hierarchical organizations can be very resistant to adaptation, because it often means *stopping* doing things you've been doing a long time, and people invested in the old ways may not like that."

An individual firm or an entire economic sector can formulate its own adaptation plan by using UKCIP's website. The Business Areas Climate Impacts Assessment Tool lists threats and opportunities that climate change presents to all parts of a business or industry's operations, from finance and logistics, work force and customers, to production of goods and delivery of services. One success story in the report features Taylerson's Malmesbury Syrups, which used to produce premium syrups for coffees and cooking. After assessing its vulnerabilities to climate change, the company concluded that the market for its products, which were linked to cold weather, would disappear within the next ten to twenty years. Taylerson's adjusted its product mix, introducing syrups for use with ice creams and other hot-weather foods. The shift was "extremely successful" with old and new clients alike, UKCIP reported.

But voluntary efforts at adaptation have definite limits, said Mark Goldthorpe, who oversees business outreach for UKCIP. "I have found that businesses are very, very good at their core, short-term concerns: doing what's necessary to get their products made, responding to customers, dealing with shareholders. They are not as good at long-term issues, especially things that are not required by regulations." By contrast, all of the UK's private water companies are implementing some form of adaptation, and Severn Trent, which provides water and sewage services to 3.7 million households in England and Wales, has a particularly aggressive program. "What differentiates the water sector is that it's regulated," Goldthorpe explained. The government's Environment Agency requires water companies "to plan twenty-five years ahead and specifically to take account of climate change," he added. "Water companies must update their plans every three years, both on business and environmental grounds, and they have to say what investments they plan to take account of climate change."

Besides government regulation, another potential source of pressure is investors, since companies need the financial capital that investors control. In recent years some large investors have come to recognize the economic risks posed by climate change, and a few have begun warning that companies need to manage these risks if they wish to remain attractive to investors. In North America, the Investor Network on Climate Risk, a group of eighty institutional investors with collective assets of $8 trillion, has been pressing this case. But the INCR group has focused overwhelmingly on mitigation, saying very little about adaptation. Meanwhile, "climate change is still not a regular agenda item for most boards [of directors]," said a 2008 report by the Carbon Disclosure Project, which surveys corporate actions on behalf of institutional investors with $64 trillion of assets (as of 2010). The project's database includes information on 2,200 companies from Europe, North and South America, Asia, and Africa. Almost none of them are active in adaptation.

Again, Britain is the exception to the rule. In 2008, three leading UK investment firms published *Managing the Unavoidable,* a report that warned that climate change "will ... leave companies with difficult business decisions; existing business models will need to change and new business opportunities will emerge. Inevitably, some existing businesses will fail." Authored by officials from Insight Investment, Universities Superannuation Scheme (which claims to be the second-largest pension fund in the UK), and Henderson Global Investors, the report urged that "the risks and opportunities presented by adaptation should be explicitly identified and integrated into overall corporate risk management and strategic planning processes." The goal of *Managing the Unavoidable*, said Rory Sullivan of Insight Investments, was "to legitimate discussion of adaptation and urge companies, investors and government to engage with it the way they already engage with mitigation."

The World's Biggest Industry

The key player to watch in all this is the insurance industry. No other industry has followed the climate issue so closely for so long; no other industry has so much to lose; and no other industry is better positioned to persuade others of the need for action.

The insurance industry is the "big exception" to the business community's general complacency about climate change, said UKCIP's Goldthorpe. "Insurance companies understand their exposure to other people's risk," he explained, "so it's in their direct business experience that weather affects them." One of the first wake-up calls came in 1992, when Hurricane Andrew and other big storms cost the global insurance industry $17 billion. Sensing an opportunity, Jeremy Leggett, a former oil industry geologist who had gone on to direct climate work at Greenpeace, began talking with insurance companies. By 1995, Swiss Re, one of the world's largest, was running full-page advertisements in the *Financial Times* arguing that "giant storms are triggered by global warming; this is caused by the greenhouse effect; which is, in turn, accelerated by man." Swiss Re and over thirty other leading companies—none of them American—signed an accord with the United Nations committing to help reduce environmental risk and address climate change. "They know that a few major disasters caused by extreme climate events ... could literally bankrupt the industry in the next decade," explained Hans Alder, director of the UN Environment Programme.

Leggett also organized insurance executives to attend UN climate negotiations in Berlin in 1995, where the executives lobbied diplomats to
impose limits on greenhouse gas emissions; it was the first time governments learned that the oil and coal industries did not speak for all businesses on the climate issue. Leggett's goal was to persuade the insurance industry to use its leverage over global capital flows to kick-start
the solar energy revolution. "The insurance industry collects some $1.4
trillion in premiums every year," Leggett told me at the time. "Much
of that $1.4 trillion is reinvested in fossil fuels, which only make things
worse, and almost none in solar and other renewables. We'd like to reverse that."

One reason insurance companies have so much leverage is that they
belong to the biggest industry in the world. Globally, the industry's
companies control an estimated $16 trillion worth of assets, a conglom-
eration of wealth greater than the gross domestic product of the United
States. As such, the industry is a leading source of the investment dollars that lubricate global capitalism. Depending on where the companies invest their money, they help determine whether climate-friendly
technologies such as solar power are encouraged and whether maladap-
tive investments such as luxury hotels on low-lying coastlines are not.
Finally, because businesses and households in the wealthy nations of the
world cannot operate without insurance, their decision makers must
heed the industry's views on which risks are insurable at what price. Just
as insurers have long refused to write policies for buildings that have
not passed a fire inspection, so in the future they may refuse (or charge
exorbitantly) to insure properties that are imperiled by rising seas, hurricanes, and other manifestations of climate change.

In recent years, a few insurance companies have moved beyond
sounding alarms about climate change to taking concrete steps against
it, though sometimes consumers have paid dearly in the bargain. More
than 500,000 homeowners in the state of Florida lost their insurance
coverage in the two years following the hurricanes of 2004 as insurers
decided that some coastal properties were too vulnerable to profitably
insure. After flooding in 2007 cost insurers $7 billion in losses, the As-
sociation of British Insurers publicly warned that large parts of Britain would become uninsurable unless the government invested more in
flood defenses. More constructive actions came from both Swiss Re and

Munich Re, when they pledged to make their operations carbon-neutral by 2012 (meaning that they will reduce their use of carbon-based fuels wherever possible and offset remaining emissions by investing in wind power or other climate-friendly steps). On the adaptation side, Tokyo Marine & Nichido said it had replanted twelve thousand acres of mangrove trees in Southeast Asia, strengthening coastal resilience to storms as well as helping the firm become carbon-neutral.

Nevertheless, such initiatives remain rare, according to Evan Mills of the U.S. Department of Energy's Lawrence Berkeley National Laboratory, perhaps the world's leading expert on the insurance industry and climate change. Each year since 2005, Mills has compiled a survey of the industry's actions on climate change for the Investor Network on Climate Risk. His 2007 report, *From Risk to Opportunity,* found that only one in ten insurers was "working in a visible way to understand the mechanics or implications of climate change." Worse, whether through ignorance or guile, most U.S. insurers were misleading outsiders about the industry's vulnerabilities to climate change. "Only 15 percent of U.S. insurers even mention climate change on their 10-K [forms filed with the government's Securities and Exchange Commission], which are supposed to describe all issues material to a company," Mills discovered. Two years later, in his 2009 report, Mills noted that many more insurance companies were now taking climate-related actions, but this increase was "only the tip of the iceberg compared with what the industry could be doing and what is needed." Insurers, he added, were doing much more on mitigation than on adaptation. Nevertheless, Leggett's dream of insurance companies financing the solar revolution remained only a dream. Mills, who occasionally consults for insurance companies, identified "$11 billion [of insurance industry investments] in renewable energy or energy efficiency" in 2008, he told me, adding that the actual number might be somewhat larger. "That's not nothing," he said, remarking that it amounted to 5 percent of total global investment in green energy in 2008. "But it's tiny" compared to the industry's assets and its vulnerabilities to climate change, he said.

The insurance industry needs to adopt a variation of "Avoid the unmanageable, manage the unavoidable," said Mills: insurance companies should strive to both increase customers' resilience and reduce climate change itself. And there are many practical, proven steps that can be taken. FM Global, one of the most profitable U.S. insurers in 2005, the year of Hurricane Katrina, claims that customers who implemented all of its resilience recommendations, such as installing storm shutters, experienced only one-eighth the losses of clients who did not, and at relatively low cost—$2.5 million of up-front costs to avoid $500 million in damages. On mitigation, insurers must lead by example, adjusting their own investment practices to foster the transition to a low-carbon economy. Munich Re, for example, has pledged that at least 80 percent of its investments will comply with strict sustainability criteria.

The rest of us, adds Mills, must also do our part. It is easy to criticize insurance companies whose main response to climate change has been to triple premiums, jack up deductibles, or drop customers. But blaming the messenger is too easy. Angels could be running insurance companies and climate change would still increase the risks and costs of extreme weather events and thus oblige companies to raise rates. "Insurance," Mills observes, "is not an entitlement. The increase in rates and deductibles is an economically appropriate signal that helps make the costs of climate change more visible to society. Higher prices should encourage all of us—governments, businesses, and individuals—to make our societies more climate-resilient."

Personally, I'm relatively optimistic that a critical mass of insurance companies and, for that matter, business enterprises in general will eventually rise to the occasion. That may sound surprising, given the record of inattention detailed in this chapter. But self-interest is a powerful motivator, and the best businesses prosper in part because they recognize and adapt to new realities even when those realities are unwelcome. As Darwin said, the species that survives is not the one that is strongest or the most intelligent—it is the one that is the most adaptable. As climate change intensifies, it will become increasingly clear to business leaders that both mitigation and adaptation are not options but necessities. The question is how many of them will come to this realization in time.

8 | How Will We Feed Ourselves?

Humans can adapt to climate change. What we don't know is how quickly we can adapt.

—DAVID LOBELL, Professor, Program on Food Security
and the Environment, Stanford University

THE TOWN WHERE Chiara has been growing up has many attractions for me, but as a former farm boy I especially appreciate the fact that we live in a bona fide agricultural community. Indeed, our town was one of the birthplaces for the organic food movement that has swept across the United States and parts of Europe in recent years. The pumpkin patch where Chiara and her cousin pick out jack-o'-lanterns every Halloween is across the road from one of the first big organic farms established in California, in 1974; the farm still supplies many of the greens served in dozens of Bay Area restaurants. And that's just one of the local organic operations. If Chiara and I walk out our door and, instead of heading down toward the beach, cross the creek and hike up the hill, in a few minutes we come to a pasture where the cattle of Niman Ranch, one of the first organic beef operations in California, graze on grass (with a stellar ocean view). Virtually all of the fruits and vegetables sold in our town's food co-op are either organic, local, or both, as are the cheeses, milk, and other dairy products.

I feel lucky that Chiara gets to eat such healthful, local food. Other parents may find it hard to believe, but Chiara truly enjoys most vegetables and fruits. (Of course, like most little kids, she also loves pasta.) It doesn't hurt that her mother is an excellent cook, or that we grow quite a few of our own vegetables. Our garden is right outside my writer's studio, and we make sure to include Chiara in the planting, harvesting, and other chores. It's good to know where your food comes from. And as the second era of global warming advances, it will be good to be able to grow some of your own food as well.

A few weeks before Chiara's fourth birthday, the new First Lady of the United States planted her own organic garden on the South Lawn of the White House. Joined by a class of local fifth-graders, Michelle Obama lifted the first shovel of dirt on a 1,100-square-foot plot that would feature fifty-five kinds of vegetables, including spinach, peppers, kale, collards, and tomatoes (but no beets—President Obama reportedly doesn't like beets). Various herbs and berries would also be grown in the garden, which would be fully visible to the thousands of tourists and other pedestrians who pass by the White House daily.

There was no doubt the First Lady was sending a message with this gesture, but the message was more subversive and far-reaching than most media coverage recognized. Michelle Obama's stated message was clearly aimed at other mothers across America: fresh food tastes better and is better for you, so both kids and grownups should eat lots more of it. "A real, delicious heirloom tomato is one of the sweetest things you'll ever eat," the First Lady told the ten-year-olds, adding that freshly picked vegetables were what had gotten her own two daughters to try new kinds of foods.

What made Mrs. Obama's message so subversive was something she left unsaid: the food most Americans are eating is neither fresh, tasty, nor healthful. Over the past fifty years, the United States has become, in author Eric Schlosser's telling phrase, a fast-food nation. What the typical American eats is not so much food as highly processed food derivatives that have traveled thousands of miles since leaving the farm, losing along the way most of the flavor and nutritional value they once possessed. To disguise such losses, food manufacturers overload products with artificial fats, salts, and sweeteners, especially corn syrup—additives that, along with the massive portions typically served in the United States, help explain why nearly one in three Americans is obese.

By publicly championing fresh local food, Mrs. Obama clearly hoped to entice Americans away from their junk food past to a healthier, more delicious future. Which is what made her message so far-reaching. Change America's eating habits and you could change the world.

The Midwest Will Bake

America's food system is the world's most advanced expression of industrial agriculture, the dominant paradigm for food production over the past fifty years. The system's great strength is its ability to produce gargantuan amounts of food. By adding large amounts of mechanization, chemicals, and irrigation to traditional farming methods, and by exploiting the economies of scale of vast, single-crop plantings, industrial agriculture exponentially increased the productivity of farming. Between 1950 and 1984, as the model spread first across the United States and then around the world, global grain production rose 260 percent—the fastest increase in human history. When combined with newly developed "miracle seeds," industrial agriculture brought the so-called Green Revolution to poor nations as well. Cereal production more than doubled in poor countries between 1961 and 1985, according to Gordon Conway, a former president of the Rockefeller Foundation, which, along with the U.S. government, was the Green Revolution's chief promoter.

Industrial agriculture's great drawback, however, has always been its ecological destructiveness. The application of chemical pesticides, herbicides, and fertilizer has deadened soil, poisoned wildlife and farm workers alike, and polluted waterways, undermining the foundations of agriculture. Runoff from midwestern farms and livestock feedlots has flowed down the Mississippi River into the Gulf of Mexico, where it has killed virtually all underwater life in a four-hundred-square-mile "dead zone" south of New Orleans. More than two hundred such dead zones now exist around the world. Meanwhile, industrial agriculture has sucked irrigation water from underground at lopsidedly unsustainable rates. Aquifers beneath some of the world's most important food production regions—the U.S. Midwest, the North China Plain, the Punjab of India—are fast approaching exhaustion.

Finally, the huge amounts of fossil fuel used to produce fertilizer, run farm equipment, and transport food to market are a major source of greenhouse gas emissions and local air pollution. Coal-fired power plants and gas-guzzling vehicles get more criticism, but factory farms, restaurants, supermarkets, and food miles—the distance a given food item travels between farm and fork—are bigger global warming culprits. The agricultural sector, including forestry, is responsible for roughly 31 percent of global greenhouse gas emissions, more than any human activity except for the constructing, heating, and cooling of buildings. Meat production alone may account for as much as 18 percent of total global emissions, according to the UN's Food and Agriculture Organization, in part because livestock emit large amounts of methane, a greenhouse gas roughly twenty times more potent than carbon dioxide.

In recent years, biofuels have been touted as a way for the agriculture sector to help reduce greenhouse gas emissions; turning plants into liquid fuels that can power vehicles could, it was said, reduce oil consumption. However, since most biofuels have relied on corn, which has very poor conversion rates, emissions reductions to date have been small. (Relying on cellulose or sugarcane, by contrast, could reduce emissions by 90 percent.) Meanwhile, diverting land to biofuels has reduced food production, driving prices up and increasing hunger. The soaring prices of wheat, rice, and other staple crops in 2007 and 2008 provoked food riots in more than a dozen countries as hungry people took to the streets. As much as 70 percent of the price increase was due to the diversion of agricultural land to biofuels, according to studies by the IMF and World Bank.

Unless fundamentally reformed, it is questionable whether industrial agriculture can survive a future shaped by climate change and peak oil. Not only does industrial agriculture exacerbate both problems; it is extremely vulnerable to them. "A key part of resilience [to climate change] is going to be diversity, and we have a terrible record over the last fifty years at fostering diversity in agriculture," explained Sara Scherr, president of Ecoagriculture Partners, a nongovernmental organization in Washington, DC. The vast monocultures favored by industrial agriculture—the miles and miles of Iowa farmland devoted solely to corn—are very susceptible to sudden stresses, be they a spike in temperature or the arrival of an unfamiliar pest or disease. Meanwhile, petroleum still powers virtually all the tractors, combines, and other machinery that help plant and harvest food, as well as the trucks and airplanes that transport it to consumers. That was fine when oil was plentiful in the twentieth century, but it invites disaster in the twenty-first. "You still see the claim that the Green Revolution saved the lives of millions of people, so there's this notion that it's the only way to do agriculture," said Fred Kirschenmann, a professor of agriculture at Iowa State University. "But nobody bothers to ask, 'If oil goes to $300 a barrel and [the supply of] water is half of today, are we still going to be able to feed the world with this system?'"

Climate change will radically alter the conditions facing farmers, generally for the worse. A few northern climes—think Canada and Russia—may benefit temporarily from warmer temperatures and longer growing seasons, scientists say, but the vast majority of projected impacts will reduce the amount, quality, and variety of food produced, making prompt and skillful adaptation essential.

The Midwest of the United States boasts some of the richest soils and most productive farmers on earth. But the scorching temperatures that are locked in over the next fifty years will make it much harder for the region's farmers to maintain current production levels, much less cope with a global population projected to grow in size and aspirations. Farmers got a taste of what lies in store in 1988, when a brutal heat wave and drought struck the Midwest, parching soil and withering crops. The United States suffered $40 billion in economic losses, the great majority of them in the midwestern farm belt. In Iowa, corn and soybean production fell by 20 to 25 percent. Many crops that did get harvested could not get to market because the Mississippi River got too dry to carry barge traffic. *The extraordinary heat of 1988 will become the norm over the next few decades*, according to U.S. government scientists, with most Iowa summers being even hotter than in 1988 (my emphasis).

Corn, the major farm crop (by volume) in the United States and the basis for most products found in U.S. supermarkets, will be especially challenged in the superheated conditions of the next fifty years. Corn does not reproduce at temperatures higher than 95°F, and yields can decline even at lower temperatures. In the past, this was not a great problem. Records from the twentieth century show that Iowa experienced three straight days of 95°F temperatures only once a decade. But by 2040, if global greenhouse gas emissions remain on their current high trajectory, Iowa will likely experience such hot spells in three summers out of four, professors Katharine Hayhoe of Texas Tech University and Donald Wuebbles of the University of Illinois have calculated.

Extreme heat also means trouble for livestock. Cattle and pigs are, like humans, warm-blooded creatures that can die if overheated. Even at temperatures of 75°F to 80°F, cows' milk production declines. "Swine, beef and milk production are all projected to decline in a warmer world," states the climate impacts report of the U.S. government's National Oceanic and Atmospheric Administration.

Winter temperatures will also be higher, bringing a range of additional problems. Many fruits need long winter chilling periods to achieve optimal growth; apples and some types of berries (especially cranberries) will suffer. At the same time, many crop diseases and pests will benefit from the warmer winters, raising the likelihood of large crop losses come spring and summer. Weeds will benefit more than food crops do from warmer temperatures and from the increased concentration of carbon dioxide in the atmosphere. As it happens, the most common herbicide in America, Monsanto's product Roundup, "loses its efficacy on weeds grown at the CO_2 levels projected to occur in the coming decades," NOAA reported. "Higher concentrations of the chemical and more frequent spraying thus will be needed, increasing economic and environmental costs. . . ."

Water, an ingredient as essential as sunlight to growing crops, will also become very problematic. As in the Netherlands, total precipitation in the Midwest is projected to remain roughly the same but arrive in more concentrated bursts. Therefore farmers can expect both more downpours and more droughts. The droughts will occur mainly in summer, the very time plants need more water to cope with higher temperatures. This scenario will be especially challenging because less than 1 percent of the region's farmland has access to irrigation.

Midwestern farmers have already experienced the havoc ill-timed downpours can wreak. In June 2008, heavy rains in Wisconsin and Iowa swelled the Mississippi River seven feet above flood level. Hundreds of thousands of acres were inundated. The floods hit just as farmers were preparing to harvest their winter wheat and plant the summer's corn and soybeans, causing $8 billion in losses. "Some farmers were put out of business, and others will be recovering for years," noted NOAA.

A Job for Superman?

Farmers and agricultural experts I've interviewed emphasize that there are ways to adapt to these and other projected climate impacts. For example, as with wine grapes, farmers could change the types and varieties of crops they plant, choosing those more tolerant of heat and aridity. Planting dates can be shifted; sowing corn earlier in the spring could lower the probability of encountering 95°F days during germination. Livestock can be protected by building sheds to provide cooling shade. Above all, farmers must improve the quality of their soil, which has been degraded by industrial agriculture's practice of applying large quantities of fertilizer, pesticides, and herbicides. Substituting compost and other sources of organic matter would increase the soil's fertility as well as its ability to retain water—crucial to surviving the hot, dry weather ahead.

But one cannot adapt to climate change without first understanding and accepting the need to adapt, and in many farming communities people still aren't convinced that man-made climate change is even real. "One approach that could be taken is the U.S. Farm Bureau's, which is to deny that climate change is happening," said Professor Wuebbles, referring to the independent, voluntary organization that has spoken to and for farmers since the 1920s. "At the Farm Bureau's annual meeting in January 2010, they had one speaker on global warming—a lawyer from a denialist think tank. I tell them, 'I'm a scientist, I'm telling you what the vast majority of scientists have concluded. Why would you believe that a lawyer is telling the truth?' The Farm Bureau is very influential, so farmers aren't getting the message they need to get. They absolutely need to be thinking about adaptation."

Perhaps nowhere is the imperative of avoiding the unmanageable and managing the unavoidable of climate change more challenging than in relation to food. Already, 1 billion people—one of every seven individuals on the planet—suffer malnutrition, meaning that they are often hungry and lack sufficient calories to live full lives. Most of the malnourished are females, many are children, nearly all are poor. Indeed, poverty is the single biggest cause of hunger. There is more than enough food produced on earth to provide everyone with adequate nutrition, but many poor people—who tend to spend more than half their incomes on food—cannot afford to buy enough to stay healthy. One result is that a grotesque amount of edible food is simply thrown away. As Tristram Stuart details in his book *Waste,* "Farmers, manufacturers, supermarkets and consumers in North America and Europe discard up to half of their food—enough to feed all the world's hungry at least three times over." Unfair global trade rules also spur hunger. Rich northern governments lavishly subsidize their own farmers, enabling them to dump wheat, corn, and other crops on the world market at prices that do not cover production costs. Farmers in poor nations cannot compete with the subsidized products, so they lose income and spiral down into poverty.

If humanity is to feed itself in the second era of global warming, we must address the food problem from (at least) two directions at once. To avoid the unmanageable, we must radically reduce the greenhouse gas emissions of the agricultural sector. That will require far-reaching changes in farming practices and technologies, which in turn implies reforming the economic mechanisms and political structures behind them. To manage the unavoidable, we must strengthen the food system's resilience, helping farmers to maintain production levels in the face of water shortages, extreme heat, and other inhospitable conditions. Vital to that effort is restoring and protecting the soils, forests, rivers, and other ecosystems that make agriculture possible in the first place.

It sounds like a job for Superman. But there are reasons for hope, and ironically, some of the best come from a continent often associated with hopelessness.

A Quiet Green Miracle

Stories that sound too good to be true usually are; an honest journalist learns that pretty early in his or her career. But every so often there is an exception. The exception I'm about to describe is from Africa, which makes it doubly welcome. For Africa is not only the continent where our species was born; it is also the continent climate change will hit the hardest.

Part of what makes Africa so vulnerable is that it is already one of the hottest, driest places on earth. The most famous desert in the world, the Sahara, occupies the northern third of the continent. Below that is the Sahel, a strip of savanna that stretches like a belt across the width of the African landmass, separating the Sahara and Sahel from the rainforests to the south. I spent five months in the eastern Sahel—Kenya, Sudan, and Uganda—while researching *Earth Odyssey,* so I was familiar with its parched, scorching climate. For this book, I went to the French-speaking western Sahel, where stepping into the midday heat was like walking into an invisible wall. The day I entered Mali, in May 2009, its famed traveler's destination of Timbuktu ranked as the hottest city in the world at 114°F (45°C).

I spent that night about a hundred miles to the south, on the Plateau Dogon, a rocky, sun-baked area known for its ancient cliff-side burial sites. The lodgings were very basic, a cross between campsite and budget hotel. I was assigned to a square concrete room with a cot, a mosquito net, and a fan that didn't work. An outdoor shower offered a brief trickle of water that was still hot from having sat in a tub on the roof all day. As I lay down to sleep, there was not even a whisper of breeze. I tossed and turned all night, unable to drop off despite my exhaustion.

Friends later assured me, back in the comfort of the United States, that the relentless heat bothered Sahel natives less because "they are used to it." There is some truth in that, though not as much as outsiders often think. The next morning, I dragged myself out of bed just after dawn, feeling grouchy and, I'm embarrassed to say, a little sorry for myself. A rooster's crow led me to peer over the wall that separated my lodgings from the property next door. A heavyset woman in a yellow T-shirt was striding across the yard, her forehead gleaming with sweat as she carried a large jug of water on her head. Despite the heat, she had clearly carried that water a considerable distance already—women in Africa often walk hundreds of yards to fetch water—and the jug had to supply all of her family's daily needs: drinking, cooking, washing. The scene reminded me that this woman and her family, like millions of other Africans, live with intense heat every day of their lives. I slapped my cheeks and told myself to stop complaining.

People in the Sahel may be more accustomed to heat, but that doesn't mean they don't suffer from it. A couple of days later in Burkina Faso, I got to know a schoolteacher named Tirouda Sectard. Tirouda looked about thirty, and he was luckier than most. He moonlighted as a radio producer, and that gave him occasional access to a studio—more of a closet, really—that had to be air-conditioned to protect the equipment. We collaborated on a radio story one day, and afterward I asked him how local people survived the heat, confessing that it sometimes left me all but unable to function.

"The same for us," he replied. "The same for us. It is too hot." Air conditioning helped, he said, but of course most locals had no access to it. "Most people suffer," he said. A pained look came over his face. "It is very bad for the old people," he added. "Old people die every day when it's this hot. The heat is too much for them."

Yet temperatures in Africa are expected to rise substantially in the years ahead and droughts worsen. According to a 2009 study coauthored by Marshall Burke, a professor at Stanford University's Program on Food Security and the Environment, six countries in the western Sahel—Mali, Burkina Faso, Niger, Sierra Leone, Senegal, and Chad—will face temperatures by 2050 that are "hotter than any year in historical experience." How peasants on the Plateau Dogon and other sun-baked areas in the western Sahel will manage this extra heat is difficult to fathom.

A second reason Africa is particularly vulnerable to climate change is that it is the poorest continent on earth. Africa has the world's largest proportion of very poor people (those earning a dollar a day or less) and its largest proportion of chronically hungry people. Most African governments are also poor, so their capacity to adapt is quite limited. The southern African nation of Malawi lacked a working barometer when a New York Times reporter visited in 2007—no surprise, since the total budget for the government weather service was a mere $160,000 a year.

So when I first heard about the phenomenon that drew me to the western Sahel—that thousands of farmers were deliberately and fairly successfully adapting to climate change, despite not knowing the term—you can see why it sounded too good to be true. I had to go see for myself. I ended up joining a trip that twenty-two activists from nongovernmental organizations in Mali and Burkina Faso were making to rural areas of their two countries. Funded by the German Catholic development group Misereor, the trip was organized by Chris Reij, a Dutch environmental specialist at VU University Amsterdam who has worked on agricultural issues in the Sahel for thirty years. Our group divided into two teams and in a week of travel covered hundreds of miles of territory, visited dozens of villages, and spoke with scores of farmers and local people.

Reporting on climate change is often a cheerless task, but in the western Sahel I observed a quiet green miracle that convinced me that all is not lost. Using simple techniques that cost them nothing, millions of small farmers throughout the region have begun protecting themselves against the scorching heat and withering drought of climate change. Their methods amount to a poor man's version of organic farming: fortifying soil with manure rather than chemical fertilizer, growing different crops on the same piece of land (known as intercropping), relying on natural predators to counter pests rather than applying pesticides. In the process, farmers in the western Sahel have rehabilitated millions of acres of degraded savanna that was on the verge of becoming desert, thus increasing the amount of land available to grow food. The transformation is so stark and pervasive that it is visible from outer space, courtesy of satellite images recorded by the U.S. government. Food yields have risen substantially; malnutrition has decreased. To be sure, the western Sahel remains a severely impoverished place where population growth is much too high and education and employment possibilities are scarce. And it is an open question how long these adaptation methods will remain effective if the outside world doesn't reverse global warming soon; every form of adaptation has its limits. But if some of the poorest farmers in the world can achieve so much, it suggests the rest of us can do even more. And talk about sounding too good to be true: the greening of the Sahel also points to a partial solution to the mitigation half of the climate challenge—to the need to slash the amount of greenhouse gases already in the atmosphere.

But that's getting ahead of the story. For the moment, suffice it to say that farmers in the western Sahel have achieved their remarkable success by deploying a secret weapon often overlooked in wealthier places: trees. Not planting trees. Growing them.

"Trees Are Like Lungs"

Yacouba Sawadogo was not sure how old he was. With a hatchet slung over his shoulder, he strode through the woods and fields of his farm with an easy grace. But up close his beard was gray, and it turned out he had great-grandchildren, so he had to be at least sixty and perhaps closer to seventy years old. That means he was born well before 1960, the year the country now known as Burkina Faso gained independence from France, which explains why he was never taught to read and write. Nor did he learn French. He spoke his tribal language, Mòoré, in a deep, unhurried rumble, occasionally punctuating sentences with a brief grunt. Yet despite his illiteracy, Yacouba Sawadogo is a pioneer of the tree-based approach to farming that has transformed the western Sahel over the last twenty years.

"Climate change is a subject I have something to say about," said Sawadogo, who unlike most local farmers had some understanding of the term. Wearing a brown cotton gown, he sat beneath acacia and zizyphus trees that shaded a pen holding guinea fowl. Two cows dozed at his feet; bleats of goats floated through the still late-afternoon air. His farm in northern Burkina Faso was large by local standards—fifty acres—and had been in his family for generations. The rest of his family abandoned it after the terrible droughts of the 1980s, when a 20 percent decline in annual rainfall slashed food production throughout the Sahel, turned vast stretches of savanna into desert, and caused millions of deaths by hunger. For Sawadogo, leaving the farm was unthinkable. "My father is buried here," he said simply. In his mind, the droughts of the 1980s marked the beginning of climate change, and he may be right: scientists are still analyzing when man-made climate change began, some dating its onset to the mid-twentieth century. In any case, Sawadogo said he had been adapting to a hotter, drier climate for twenty years now.

"In the drought years, people found themselves in such a terrible situation they had to think in new ways," said Sawadogo, who prided himself on being an innovator. For example, it was a long-standing practice among local farmers to dig what they called *zai*—shallow pits that collected and concentrated scarce rainfall onto the roots of crops. Sawadogo increased the size of his *zai* in hopes of capturing more rainfall. But his most important innovation, he said, was to add manure to the *zai* during the dry season, a practice his peers derided as wasteful.

Sawadogo's experiments proved out: crop yields duly increased. But the most important result was one he hadn't anticipated: trees began to sprout amid his rows of millet and sorghum, thanks to seeds contained in the manure. As one growing season followed another, it became apparent that the trees—now a few feet high—were further increasing his yields of millet and sorghum while also restoring the degraded soil's vitality. "Since I began this technique of rehabilitating degraded land, my family has enjoyed food security in good years and bad," Sawadogo told me.

Chris Reij and other scientists who have studied the technique say that mixing trees and crops—a practice they have named "farmer-managed natural regeneration," or FMNR, and that is known generally as agro-forestry—brings a range of benefits. The trees' shade and bulk offer crops relief from the overwhelming heat and gusting winds. "In the past, farmers sometimes had to sow their fields three, four, or five times because wind-blown sand would cover or destroy seedlings," said Reij, a silver haired Dutchman with the zeal of a missionary. "With trees to buffer the wind and anchor the soil, farmers need sow only once." Leaves serve other purposes. After they fall to the ground, they act as mulch, boosting soil fertility; they also provide fodder for livestock in a season when little other food is available. In emergencies, people too can eat the leaves to avoid starvation.

The improved planting pits developed by Sawadogo and other simple water-harvesting techniques have enabled more water to infiltrate the soil. Amazingly, underground water tables that plummeted after the droughts of the 1980s had now begun recharging. "In the 1980s, water tables on the Central Plateau of Burkina Faso were falling by an average of one meter a year," Reij said. "Since FMNR and the water-harvesting techniques began to take hold in the late 1980s, water tables in many villages have risen by at least five meters, despite a growing population." Some analysts attributed the rise in water tables to an increase in rainfall that occurred beginning in 1994, Reij added, "but that doesn't make sense—the water tables began rising well before that." Studies have documented the same phenomenon in some villages in Niger, where extensive water-harvesting measures helped raise water tables by fifteen meters between the early 1990s and 2005.

Over time, Sawadogo grew more and more enamored of trees, until now his land looked less like a farm than a forest, albeit a forest composed of trees that, to my California eyes, often looked rather thin and patchy. Trees can be harvested—their branches pruned and sold—and then they grow back, and their benefits for the soil make it easier for additional trees to grow. "The more trees you have, the more you get," Sawadogo explained. Wood is the main energy source in rural Africa, and as his tree cover expanded, Sawadogo sold wood for cooking, furniture making, and construction, thus increasing and diversifying his income—a key adaptation tactic. Trees, he says, are also a source of natural medicines, no small advantage in an area where modern health care is scarce and expensive.

"I think trees are at least a partial answer to climate change, and I've tried to share this information with others," Sawadogo added. "I've used my motorbike to visit about a hundred villages, and others have come to visit me and learn. I must say, I'm very proud these ideas are spreading." In November 2009 Sawadogo flew to Amsterdam and Washington, DC, to address conferences on climate change, food security, and poverty. His message was the same one he gave me: "My conviction, based on personal experience, is that trees are like lungs. If we do not protect them, and increase their numbers, it will be the end of the world."

The Largest Environmental Transformation in Africa

Sawadogo was not an anomaly. In Mali, the practice of growing trees amid rows of cropland seemed to be everywhere. A bone-jarring three-hour drive from the Burkina Faso border brought us to the village of Sokoura. By global standards, Sokoura was very poor. Houses were made of sticks covered by mud. There was no electricity or running water. Children wore dirty, torn clothes, and more than a few were naked, their distended bellies hinting at insufficient diets. When one of our team let an empty plastic bottle fall to the ground, kids wrestled for it as if it were gold. Yet to hear locals tell it, life was improving in Sokoura.

It was a five-minute walk from the village to the land of Omar Guindo. Missing a front tooth and wearing a black smock over green slacks, Guindo said that ten years ago he began taking advice from Sahel Eco, a Malian NGO that promotes agro-forestry. Now, Guindo's land was dotted with trees, one every five meters or so. Most were young, with such spindly branches that they resembled bushes more than trees, but there were also a few specimens with trunks the width of fire hydrants. We sat beneath a large tree known as the "Apple of the Sahel," whose twigs sported inch-long thorns. The soil was sandy in both color and consistency—not a farmer's ideal—but water availability and crop yields had increased substantially. "Before, this field couldn't fill even one granary," he said. "Now, it fills one granary and half of another"—roughly a 50 percent increase in production.

Back in the village, we examined the granaries, which were built by layering mud over stick frames. Oblong in shape, the structures had sides that were six feet wide and fifteen feet tall. A notched tree trunk served as a ladder to an opening near the top. Reij was the first to climb, serenaded by jovial laughter from the crowd below; it was not often these villagers got to see a white man make a spectacle of himself. Reij played to the crowd, joking about being too clumsy to manage such a steep ladder and asking one of the grannies to help him. After inspecting all four granaries, the Dutchman descended, turned to me, and exclaimed, "This is thrilling." Pointing to the closest granary, he said, "This one still has a little millet in it. The next one is more than half full, the third is totally full, and the last is a third full. What that means is, this farmer has tremendous food security. It is now May. Harvest will be in November. So he has plenty to last his family until then and even some in reserve."

As word of such successes travels, FMNR has spread throughout the region, according to Salif Ali, a neighboring farmer. "Twenty years ago, after the drought, our situation here was quite desperate, but now we live much better," he said. "Before, most families had only one granary each. Now, they have three or four, though the land they cultivate has not increased. And we have more livestock as well."

After extolling the many benefits trees have provided—shade, livestock fodder, drought protection, firewood, even the return of hares and other small wildlife—Salif was asked by one member of our group, almost in disbelief, "Can we find anyone around here who doesn't practice this type of agro-forestry?"

"Good luck," he replied. "Nowadays, everyone does it this way."

Perhaps I should reemphasize here that these farmers were not planting these trees, as Nobel Prize–winning activist Wangari Maathai has promoted in Kenya. Planting trees is much too expensive and risky for poor farmers, Reij said, adding, "Studies in the western Sahel have found that 80 percent of planted trees die within a year or two." By contrast, trees that sprout naturally are native species and more resilient. And, of course, such trees cost the farmers nothing.

Even naturally sprouting trees were off-limits to farmers until laws were changed to recognize their property rights. Tree management was traditionally part of normal agricultural practice here, Salif explained; it was encouraged by the Barahogon, a voluntary association of farmers to which both Salif and his father belonged. But the practice was largely abandoned after first colonial and later African governments declared that all trees belonged to the state, a policy that gave officials the opportunity to sell timber rights to business people. Under this system, farmers were punished if they were caught cutting trees, so to avoid hassles they often uprooted seedlings as soon as they sprouted. In the early 1990s, a new Malian government, mindful that forestry agency officials had been killed in some villages by farmers furious about illegal burning of trees by forestry agents, passed a law giving farmers legal ownership of trees on their land (though farmers did not hear about the law until NGOs mounted a campaign to inform them via radio and word of mouth). Since then, FMNR has spread rapidly. Recently, farmers even shared their knowledge with officials visiting from Burkina Faso—twenty mayors and provincial directors of agricultural and environmental agencies. "They seemed astonished to hear our story and see the evidence," Salif recalled. "They asked, 'Is this really possible?'"

Recognizing farmers' property rights was equally crucial in Niger, according to Tony Rinaudo, an Australian missionary and development worker who was one of the original champions of FMNR. "The great thing about FMNR is that it's free for farmers," Rinaudo told me. "They stop seeing trees as weeds and start seeing them as assets." But only if they're not penalized for doing so. In Niger, said Rinaudo, FMNR had a hard time gaining traction until he and others convinced government officials to suspend enforcement of the regulations against cutting trees. "Once farmers felt they owned the trees in their fields, FMNR took off," Rinaudo recalled.

The pattern has been the same throughout the western Sahel: FMNR has spread largely by itself, from farmer to farmer and village to village, as people see the results with their own eyes and move to adopt the practice. Not until Gray Tappan of the U.S. Geological Survey compared aerial photos from 1975 with satellite images of the same region in 2005 was it apparent just how widespread FMNR had become: one could discern the border between Niger and Nigeria from outer space. On the Niger side, where farmers were allowed to own trees and FMNR was commonplace, there was abundant tree cover; but in Nigeria, the land was barren. Reij, Rinaudo, and other FMNR advocates were surprised by the satellite evidence; they had had no idea so many farmers in so many places had grown so many trees.

"This is probably the largest positive environmental transformation in the Sahel and perhaps in all of Africa," said Reij. Combining the satellite evidence with ground surveys and anecdotal evidence, Reij estimated that in Niger alone farmers had grown 200 million trees and rehabilitated 12.5 million acres of land. "Many people believe the Sahel is nothing but doom and gloom, and I could tell lots of doom-and-gloom stories myself," he said. "But many farmers in the Sahel are better off now than they were thirty years ago because of the agro-forestry innovations they have made."

What makes FMNR so empowering—and sustainable—Reij added, is that Africans themselves own the technology, which is simply the knowledge that nurturing trees alongside one's crops brings many benefits. Thus FMNR's success does not depend on large donations from foreign governments or humanitarian groups—donations that often do not materialize or can be withdrawn when money gets tight. This is one reason Reij sees FMNR as superior to the Millennium Villages model promoted by Jeffrey Sachs, the economist who directs Columbia University's Earth Institute. The Millennium Villages program focuses on twelve villages in various parts of Africa, providing them free of charge with what are said to be the building blocks of development: modern seeds and fertilizer, boreholes for clean water, health clinics. "If you read their website, tears come to your eyes," said Reij. "It's beautiful, their vision of ending hunger in Africa. The problem is, it can only work temporarily for a small number of selected villages. Millennium Villages require continuing external inputs—not just fertilizer and other technology, but the money to pay for them—and that is not a sustainable solution. It's hard to imagine the outside world providing free or subsidized fertilizer and boreholes to every African village that needs them."

Outsiders do have a role to play, however. Overseas governments and NGOs can encourage the necessary policy changes by African governments, such as granting farmers ownership of trees. And they can fund, at very low cost, the grassroots information sharing that has spread FMNR so effectively in the western Sahel. Although farmers have done the most to alert peers to FMNR's benefits, crucial assistance has come from a handful of activists like Reij and Rinaudo and NGOs such as Sahel-Eco and World Vision Australia. These advocates now hope to encourage the adoption of FMNR in other African countries through an initiative called "Re-greening the Sahel," said Reij.

"Before this trip, I always thought about what external inputs were required to increase food production," Gabriel Coulibaly said at a debriefing session after our fact-finding expedition. Coulibaly, a Malian who worked as a consultant to the European Union and other international organizations, added, "But now I see that farmers can create solutions themselves, and *that* is what will make those solutions sustainable. Farmers manage this technology, so no one can take it away from them." After a string of similar comments from other activists—"The farmers understand why they are doing this, so they will defend it," one said—Reij leaned over and, his eyes shining, whispered, "They have been transformed into FMNR champions."

Now a final fact that sounds too neat to be true: the debriefing session took place in a building on Rue de Copenhagen—Copenhagen Street. In seven months' time, the world's governments would assemble in Copenhagen for what was being described as the most important conference in the history of the climate issue. As I listened to Reij and the western Sahel activists compare notes, I wished the diplomats who would be coming to Copenhagen could have joined us and heard their enthusiasm. If humanity is to avoid the unmanageable and manage the unavoidable of climate change, we must pursue the best options available. FMNR certainly seems to be one of them, at least for the poorest members of the human family. "Let's look at what's already been achieved in Africa and build on that," urged Reij. "In the end, what happens in Africa will depend on what Africans do, so they must own the process. For our part, we must realize that farmers in Africa know a lot, so there are things we can learn from them as well."

Ecological versus Industrial Agriculture

It is not only sympathy but self-interest that causes me to linger on this story from Africa. After all, places like the Sahel are experiencing today the kind of weather the rest of us will increasingly face tomorrow. "Under climate change, a much higher proportion of [the earth's] total agricultural land area is going to function like the higher-vulnerability areas have functioned for a long time," said Sara Scherr, the president of Ecoagriculture Partners. "There are very few examples where there has been an *intention* to adapt [to climate change] but lots of places where they have *had* to adapt, and there are lessons to be learned from them. Africans have been dealing with this problem for a long time; there just hasn't been much attention paid. If we went around the world and found the ten projects in every place that are working and tried to bring them to scale regionally or nationally, that would take us a long way."

The western Sahel's lessons are especially apt for farmers in the U.S. Midwest, said Jerry Hatfield, the director of the U.S. Department of Agriculture's National Laboratory for Agriculture and the Environment. "I tell producers, 'You need to get your soil into a condition where it can store as much water as possible,'" Hatfield told me. "Soil is made of sand, silt, and clay. Organic matter—decomposed plant material, bugs, microbes—is the glue that holds soil together and helps it retain rainfall. Gardeners know this. If you return compost to your soil, it makes the soil dark and rich, easy to work with, and able to stay moist even in dry periods."

Hatfield's answer seemed to suggest that organic farming is superior to industrial agriculture as a response to climate change, but the truth, he said, is more complicated. "Climate change will lead to increased pressure from insects, diseases, and weeds," he explained. "People don't talk about this much, but it is potentially huge. These indirect impacts of climate change will affect yields as much as the direct impacts [e.g., higher temperatures and variable rainfall] do. I respect organic agriculture, but it may have less flexibility in this regard. Industrial agriculture will have to be more judicious, too; you can't just spray this problem away.

"We're going to have to rethink our entire agronomic system to determine how to deal with climate change," Hatfield argued. In the second era of global warming, organic agriculture has clear ecological advantages. The industrial agricultural system, as currently constituted, is a climate killer whose emissions must be slashed. At the same time, industrial agriculture's preference for monocultures makes the system far less resilient to most climate impacts than more diverse ecological systems are. Still, ecology is not the only issue that matters. On a planet where 1 billion people already don't get enough to eat and the population is projected to reach 9 billion by 2050, output and price also matter a great deal.

Proponents of industrial agriculture contend that only their system can produce enough food to feed the world at prices ordinary people can afford. The late Norman Borlaug, who developed the high-yield seeds of the Green Revolution, used to call environmentalists elitists whose affluence allowed them to worry about pollution rather than starvation. "If they lived just one month amid the misery of the developing world, as I have for fifty years, they'd be crying out for tractors and fertilizer and irrigation canals . . . ," Borlaug once said.

Critics counter that the Green Revolution should not be credited with reducing starvation because, despite massive production increases, plenty of hunger remains in poor countries. But that argument is disingenuous, for it ignores the fact that there would have been even more hunger without the production increases Borlaug helped set in motion. As environmentally costly as it has been, the industrial agricultural model commands a certain respect; without it, millions more would have suffered and perished from hunger over the past fifty years.

Critics of industrial agriculture are on stronger ground when they argue that the food it produces is deceptively cheap: it costs less at the supermarket, but only because governments provide industrial agriculture with all sorts of subsidies, ranging from direct payments to farmers to technical and marketing support from research agencies and the like. What's more, the supermarket price does not include the health, environmental, and social costs of industrial food: the rise of obesity; the pollution of air, soil, and water; the elimination of family farmers and the rural communities that depend on them. At the moment, these costs are borne by society as a whole. If the producers and consumers of industrial food instead had to pay these costs themselves, the supermarket price of industrial food would be much higher. But there is nothing inherently wrong with government subsidies of the food system; virtually every country in the world engages in some form of them. Subsidies merely reflect the policy decisions and underlying value judgments a society chooses to embrace. If industrial agriculture seems likely to do a better job of feeding people in the years ahead, why shouldn't governments support it? The same holds true, of course, for organic agriculture.

The real question seems to be this: If organic agriculture is truly a more climate-friendly and climate-resilient way of producing our food, can it do so without sacrificing total output?

The answer, it turns out, is different in different places. In Africa, the case for organic farming is quite strong. In 2008, a major study conducted by the UN Environment Programme and the UN Conference on Trade and Development concluded that organic agriculture "can be more conducive to food security in Africa than most conventional production systems, and . . . it is more likely to be sustainable in the long term." The study examined 114 cases of organic farming in various parts of Africa, most involving small farmers. Yields did not fall when land was converted from conventional to organic farming, and over time they increased to match the yields of conventional systems. In 93 percent of the cases, a shift to organic farming also brought environmental benefits that will be valuable in the face of climate change, including better soil fertility, water supply, flood control, and biodiversity.

But Africa is a special case. For reasons both physical and political, the continent was largely bypassed by the Green Revolution, so its conventional crop yields today remain low by global standards. Thus it is easier for ecological agriculture to match those yields. But for the many countries where industrial agriculture has become the norm, the expectation of policymakers and ordinary people alike is that the huge production volumes of the past will continue in the future. Can that expectation be met if these countries shift a significant part of their agricultural sectors to organic? Or will the goal of increasing production trump the need to cope with climate change?

This question is now being debated by farmers, scientists, government officials, activists, and agricultural experts the world over, and the debate seems bound to intensify in the years ahead. Nowhere will it carry greater consequences than in China, the rising power of the twenty-first century and the home of one out of every five humans on earth.

"We Don't Know the Answer Yet"

Ni-hao, the one Mandarin phrase most visitors to China master, is the common greeting in China, like saying "Hello" in English. But it wasn't too long ago that Chinese people greeted one another by asking the Mandarin equivalent of "Have you eaten?" Many middle-aged and older Chinese retain vivid memories of the hungry years of the 1960s, when Mao's Great Leap Forward and later his Cultural Revolution threw the nation's agricultural system into chaos. The resulting scarcity, as journalist Jasper Becker documents in horrifying detail in his book *Hungry Ghosts,* was more a function of political hysteria than of production shortages. Granaries often had plenty of food, but Mao ordered it not distributed to the peasantry, whom he suspected of hoarding for "counterrevolutionary" purposes. Becker estimates that at least 30 million perished in arguably the greatest act of mass murder in the twentieth century.

Food security has been an issue of paramount concern to the Chinese Communist Party ever since. Beginning in 1979, after Mao's death, "the first and most important edict the Party issued [each year] concerned agriculture," Becker reports. As limited private farming was permitted, grain production increased by 42 percent between 1976 and 1984, helping to lift an estimated 200 million peasants out of severe poverty. But the rapid growth of China's massive population, along with people's desire to eat more meat and dairy products, meant that production had to keep rising—no small challenge.

Now, climate change is complicating the challenge of feeding the most populous nation on earth. "The government wants yields to remain stable or increase every year, but that is very difficult to do because of droughts and other weather extremes climate change is bringing," said Lin Erda, who has represented his country on the IPCC since the *First Assessment Report* in 1990.

In 2008, Lin served as the lead author of a separate report, *Climate Change and Food Security in China,* that was nothing short of extraordinary. Part of what made it so was its contents: the report concluded that China's "basic food supplies will become insufficient around 2030" if global greenhouse gas emissions remain high and that "overall food production will fall by 14 to 23 percent by 2050 from 2000 [levels]" if no adaptation measures are taken. Coming from one of China's top climate scientists, those were sobering words. Almost as remarkable, the report was produced with Greenpeace China, an organization the Chinese government did not exactly admire. The final surprise in the report was how sharply it appeared to diverge from current government policies. On the basis of field studies from various regions in China, the report recommended that the best way to cope with climate change was to promote "ecological agriculture." By ecological agriculture the report meant such practices as using fewer chemical fertilizers, raising ducks and fish in rice paddies (the fish decrease methane emissions; the ducks control pests), and mixing farmland and forest in ways that bear a striking resemblance to what I observed in the Sahel. Ecological agriculture, the report said, would "reduce greenhouse gas emissions from agriculture while ensuring production yield. . . . Also, due to its characteristics, ecological agriculture can more easily adapt to climate change and its associated problems such as rising temperatures, extreme weather, soil degradation, and increasing frequency of disease and pest outbreaks."

As in the rest of the world, however, doubts have been raised in China about the economics of ecological agriculture—about its effect on prices and output. Vaclav Smil, a professor at the University of Manitoba and one of the world's leading experts on China's environment, scoffed that "ecological agriculture" was "such a fashionable, empty term." If it referred to "old-fashioned organic recycling and multicrop rotations," Smil added, "those practices (as desirable as they may be) do not sit well with mass agriculture geared to supply China's mega-cities." When I relayed that critique to Ma Shiming, Lin Erda's colleague at the Chinese Academy of Sciences and a coauthor of the *Climate Change and*

Food Security in China report, he did not flinch. "That's the key question," Ma agreed, "and we don't know the answer yet. More and more research suggests that organic farming can feed the world, but whether it can feed China we can't see yet. Most research on organic farming has been done in foreign countries. In Germany, for example, yields decreased by 25 percent the first year [after the switch to organic], but then they increased and after five to seven years they equaled conventional yields. So we say to our government, we must do research in China to see if ecological agriculture can work, both for food security and for adaptation to climate change."

"Now We Use Lots of Chemical Fertilizer"

When I headed to the countryside to see examples of ecological agriculture and hear what China's farmers themselves thought about climate change, I immediately came face-to-face with one of the biggest challenges to food security in China. I had decided to visit two of the country's most important food-producing areas: Shandong, the coastal province southeast of Beijing, and Henan, its neighbor to the west, which contains some of the oldest agricultural land in the world—the flatlands watered by the Yellow River. Leaving Beijing, the train propelled me past scenery that changed scarcely at all from the city center through the distant exurbs. I passed mile after mile of high-rise apartment buildings, many already built, many still under construction. Like many other Chinese cities, Beijing was expanding outward, paving over farmland in a desperate effort to house the nation's gargantuan population.

Proceeding south past Tianjin, a city with a population (officially) of 11 million, I witnessed a second challenge to food security. The land was agricultural now: fields of green winter wheat stretched to the horizon, alternating with vast conglomerations of plastic-sheeted greenhouses where vegetables were grown. But what little unplanted land I saw was dry, dry, dry. So where did the water to grow all this agricultural bounty come from? From the many irrigation canals that stretched across the landscape, shimmering in the afternoon sun like giant silver ribbons. The problem was, these canals were fed by underground water tables that were rapidly being depleted. Scientists have estimated that the aquifers beneath the North China Plain—which is home to more than 200 million people and produces 60 percent of China's wheat—will dry up entirely in thirty years. "The rate of decline is very clear, very well documented," Richard Evans, a hydrologist who had consulted for China's Ministry of Water Resources, told journalist Susanne Wong. "They will run out of groundwater if the current rate continues."

In Shandong I met one of China's keenest promoters of ecological agriculture, Jiang Gaoming. Bright-eyed, solidly built, quick to smile, Jiang was a young-looking forty-three. He made his living as a researcher at the Institute of Botany of the Chinese Academy of Sciences and a professor at the Shandong Agricultural University in the provincial capital. But Jiang had been raised in a typical rural village, and he still spent lots of time there, so he had an intimate understanding of how peasants view the world. As we drove to his village, Jiang noted, not entirely proudly, that Shandong was China's number-one province for total agricultural production: it was the leading producer (by volume) of pork, chicken, beef, milk, vegetables, and marine products, and it was the number-two producer of wheat and corn. "But this level of production is only possible because of very intense production methods, including incredible amounts of chemical fertilizers and pesticides," he added. "Chinese farmers use two to three times more fertilizer than the world average. They apply so many chemicals that the [level of] organic matter in our soil is less than 1 percent. The normal level in the U.S. is as high as 12 percent. Much of the land here doesn't even have worms anymore—there's nothing for them to eat. They can't eat chemicals."

"But China's farmers have centuries of experience with organic fertilizer," I said. "Why are they using so many chemicals now?"

"The government subsidizes it," Jiang replied. "And farmers have used chemical fertilizers so long that the soil fertility has declined. So now the soil needs even more fertilizer to maintain yields. Adding more fertilizer reduces soil fertility even more, so it's a vicious circle."

I heard the same comment about diminishing returns when I talked with farmers in Jiang's old village, which went by the tongue-twisting name of Jiangjiazhuang. The village's former party secretary Jiang Guangchen (no relation to Jiang Gaoming) was a fifty-nine-year-old farmer with tobacco-stained teeth and callused hands, the left one of which was missing half an index finger. He had been party secretary

from 1974 to 1989, the years when market policies began supplanting
Maoist doctrine in the countryside. When I asked what was the biggest
difference between how the village farmed then and how it farmed now,
he replied, "Now we use lots of chemical fertilizer. Before, we used human and animal waste."

"Which way is better?"

"Personally, I think the old way is better. It makes the soil softer and
healthier. Production levels are about the same. In 1978, our yield was
1,050 kilograms per *mu*. Nowadays, yield is about 1,100 kilograms per
mu." The *mu* is the traditional measurement of land in China. Six *mu*
equal one acre; fifteen *mu* equal one hectare. "Farmers use chemical
fertilizer today because it is more convenient and they can see instant
effects."

"That's right," Jiang interjected. "Using organic fertilizer requires
more time and effort, whereas chemical fertilizer leaves farmers time to
go to the city and work another job. All over China, the able-bodied men
are leaving the countryside to work in cities. Almost the only people left
in the villages now are the old, the sick, and the disabled. Even many
women go. So farmers want the easy way to apply fertilizer."

Hoping to demonstrate that a more ecological approach would yield
better results, Jiang had established an eco-farm near his old village.
Staffed by graduate students from China and foreign countries, the
farm covered forty *mu* and produced a wide range of products, from
wheat and potatoes to cattle, fish, and grasshoppers. My eyes must have
given away my surprise at the grasshoppers, for Jiang smiled and said,
"You fry them with oil and salt, maybe pepper. Tasty!" As we ducked inside a small, netted enclosure that housed the insects, he added, "The
production from this enclosure alone brought in RMB 4,000 [$500] last
year in less than one hundred square meters."

Chickens on the eco-farm were let loose to wander in the cornfield,
where their appetite for insects allowed for elimination of chemical
pesticides. Cattle were fed corn stalks that would otherwise have been
burned as waste, converting a source of greenhouse gas emissions into
animal feed. The cows' manure was used as fertilizer and fuel. Stuffed
into biogas plants—underground tanks about the size of garbage
cans—the manure was processed into the gas that powered lights and

computers in the dormitory. "These are closed-loop systems," Jiang said proudly. "Nothing is wasted."

The most important experiment was taking place in front of the dormitory, where the students were growing winter wheat, the irreplaceable ingredient in the noodle dishes of northern China. All chemical additives were banned from the plot; the only enhancements came from cow and chicken manure. "Last year was the first year of the experiment, and the yield was very low," Jiang said, "I think partly because this plot was covered with plastic during the construction of the dormitory. But this year we expect the yield to be almost equal to conventional [agricultural] methods."

If Jiang is right and ecological agriculture's yields do match those of conventional agriculture, he would seem to have a persuasive case to make to Chinese authorities. Equally significant, he claimed his eco-farm had cracked the other economic challenge: profitability. "This year, this farm made RMB 6,500 [$812.50] per *mu*. Compare that to the RMB 500 per *mu* that conventional farmers make growing grain, or even the RMB 3,000 per *mu* they make with watermelon."

But—and it's a significant but—Jiang conceded that the economic benefits of organic farming do not come immediately: "The main obstacle to ecological farming is that it takes three to five years to make the transition from the conventional way of doing things. Most farmers don't want to wait that long. They can't. They are already on the edge of survival."

Industrial Agriculture with Chinese Characteristics

China is as focused on agriculture as any government in the world, but despite the urgings of scientists like Lin and scholar-activists like Jiang, much of official policy remains wedded to the old way of doing things. It seems the government is betting the nation's food security on an intensification of current practices—what amounts to "industrial agriculture with Chinese characteristics."

China cannot reap the productivity benefits of large farms because that would require throwing millions of peasants off their small landholdings; since land is a form of social security in rural China, such a policy would risk widespread suffering, anger, and social unrest. Nor can China expand the total land under cultivation, given the country's increasing population and pell-mell development. Thus the party hopes to increase the productivity of existing farms. This it will do, premier Wen Jiabao indicated at the 2009 party congress, by applying greater amounts of the traditional inputs of industrial agriculture—more fertilizers, pesticides, and irrigation—while also slightly raising the guaranteed government price for grain, boosting farmers' incentive to grow it.

This strategy is by no means assured of success. The diminishing returns some farmers are already experiencing with chemical fertilizers are a warning sign, and the diminishing supply of water poses an even bigger threat. The government has said it wants to increase the amount of irrigated land in China by 20 percent, but where will the additional water come from? The North China Plain aquifer is sprinting toward exhaustion. When I visited China in May 2009, its north had just suffered the worst drought in fifty years, and Chinese scientists project that climate change will reduce overall rainfall in the decades ahead. Meanwhile, rapid economic development and weak environmental enforcement have given China some of the most polluted water supplies on earth, "to the point that vast stretches of rivers are dead and dying, lakes are cesspools of waste, groundwater aquifers are over-pumped and unsustainably consumed, uncounted species of aquatic life have been driven to extinction and direct adverse impacts on both human and ecosystem health are widespread and growing," as Peter Gleick wrote in *The World's Water 2008–2009*.

About the only bright spot in China's water outlook is that the melting of the Himalayan snowpack may—I emphasize *may*—not have the dire consequences often supposed. Numerous analysts have observed that the gradual melting of the Himalayan snowpack over the coming decades will imperil rivers on which hundreds of millions of people rely for water; I said so myself earlier in this book. But the picture is more complicated than that, for snowmelt is not the only source of water for those rivers; rain also plays a greater or lesser role, depending on the particular river. For example, snowmelt is apparently only a minor factor in the flow of the Yangtze, China's largest river. "Snowmelt provides only 3 percent of China's fresh water," Ding Yihui, the former head of the China Meteorological Administration and one of the nation's top climate experts, told me. "Precipitation accounts for 97 percent. So if snowmelt changes, it affects only 3 percent of the total supply."

The government absolutely needs the Yangtze to keep flowing, for it is key to its proposed solution to China's water shortages: the South-to-North Water Transfer Project. This enormous, three-pronged project is designed to divert 45 billion cubic meters of water a year from the Yangtze River basin to the parched north. The idea is based on a basic fact about China's water: the bottom half of the country, from the Yangtze southward, has more than enough water, while the northern half is one of the driest inhabited areas on earth. The notion of reengineering China's hydrology dates back to the faith in mega-projects that Mao shared with his Communist brothers in the Soviet Union: science and technology would bend Nature to man's will. The project's current design envisions an eastern route that will extract water from the lower Yangtze and pump it north through a 1,200-kilometer-long canal; a middle route that will send water from the middle of the Yangtze basin northward; and a western route that will inject water from the upper reaches of the basin to the severely depleted Yellow River. The eastern route has been completed, but problems have arisen: the water being transferred is so polluted that it requires extensive treatment before it can be used even for irrigation. Environmentalists and scientists have raised other concerns, including a fear that diverting flow from the Yangtze will further erode the quality of the river and devastate its ecosystems. But the government is determined to move forward, and even as ecologically minded a scientist as Lin Erda supports the project. "Initially, when the government developed this idea, they didn't think of it as adaptation to climate change," Lin told me. "But it turns out to be an excellent example."

Improving water efficiency is a much better solution, other Chinese scientists have argued, though it has gotten less attention and support. Agriculture accounts for 70 percent of North China's water consumption, in line with the global average. But water use efficiency—the amount of water that reaches crops—is only about 40 percent, half of what U.S. farms achieve. Likewise, Chinese industry uses four to ten times more water than its counterparts in industrial countries. So the potential for efficiency savings is vast, if policies can be reformed and proper technologies introduced.

Smart efficiency measures can save farmers money, as I saw when I visited farms north of Beijing. Farmers could reduce water use by more than 50 percent by installing a drip irrigation system developed by Professor Kang Yuehn of the Institute of Geographic Sciences and Natural Resources Research of the Chinese Academy of Sciences. "Most drip irrigation systems have been developed by the U.S. or Israel, and they are good systems, but they are not suitable for China because Chinese farmers aren't educated enough to know how to operate them," Kang told me. "I developed a system where all they have to do is look at a meter and push a switch—very easy."

I saw Kang's system in operation at two government-supported research farms, but the best testimony came from a private farmer who bought the system of her own accord. Born in 1964, Zhang Guizhi was farming near the same plot of land her parents had worked. But while they had earned pittances growing wheat, she was growing table grapes and making "much more money," about 5,000 yuan per *mu* (roughly equivalent to what Jiang Gaoming said his much larger eco-farm in Shandong made). Describing herself as a "firm believer in science and technology," Ms. Zhang said she heard about drip irrigation from the local water resources bureau and installed the system in 2006. She showed me the main piece of equipment: a gray metal box about the size of a carry-on suitcase, which was connected to a series of plastic irrigation pipes that snaked along the ground beneath her rows of vines. "We are very happy with it," she said. "It has increased yields dramatically. It has also reduced our water use by about 50 percent, which means more money for us because we don't need to buy so much electricity."

But government support is essential if drip irrigation technology is to spread to less prosperous farmers, said Professor Kang. "If farmers can recover their investment in one year, they will adopt new technologies, but if it takes longer, they won't."

In sum, China has the tools and talents needed to shift its agricultural system toward a more ecological model that promises—after a brief transition period—both to enhance food security and to better cope with climate change. Making such a shift will require substantial policy reforms, however. If the findings of scientists like Kang, Jiang, Ma, and Lin are accurate—and one of the first tasks is to fund more research to test their claims—the main obstacle to shifting to ecological agriculture is the economic lag time. Kang says it takes two to three years before the investment in drip irrigation pays for itself; Jiang estimates three to five years for the shift to ecological farming; Ma cites estimates from Europe of five to seven years. More research should provide more precise numbers, but the political imperative will remain the same: if China is to embrace ecological agriculture, the government will have to subsidize the transition.

The government has been doing exactly that in regard to energy. Thanks to well-funded government directives, China has closed hundreds of the ancient coal and cement plants that were fouling the air during my visit in 1996. China has become the world's leading producer of solar panels, it has made great strides with wind power, and it is accelerating its impressive record on energy efficiency. The problems of agriculture are somewhat different, but a similar shift in policies might well produce similar results, if China's leadership is so inclined.

Are Genetically Modified Seeds the Answer?

The wildcard in food's future under climate change, not only in China but around the world, is genetic engineering. I was told during my May 2009 visit to China that the government had just launched what Ma Shiming called "a giant exploratory research project" to decide whether genetic engineering made sense. "China's interest in GMOs [genetically modified organisms] has gone up and down in recent years," Ma told me. "We are not as positive about GMOs as the U.S. and not as negative as Europe. The government doesn't want to put GMOs into the food system too quickly. It feels China's research is not yet advanced enough to make this decision. We don't know enough about GMOs' effects on human health."

Virtually every agricultural scientist I interviewed in China expressed similar caution. "The government has put a huge amount of R and D money into GMOs, but it has approved it for only six crops so far because of caution about the effects on human health," said Zhu Lizhi, a scientist who directs the Institute of Agricultural Economics of the Chinese Academy of Agricultural Sciences. Approved crops include cotton, sweet pepper, and papaya. "These are not the core of the Chinese diet, so I don't worry too much about them," Zhu added. "But personally, I'd rather not choose GMOs. I have two children, and I worry GMOs might not be too good for their genetic makeup."

No one could provide me figures on how much China was spending on its GMO research, but the tone of most comments suggested that it was considerably more than is being devoted to ecological agriculture. Again, Lin Erda is a valuable barometer. As someone with direct access to party leaders and a personal commitment to ecological agriculture, he characterized the government's position on food GMOs as "active but cautious." His personal opinion, Lin added, was that "current technologies can handle the problem through 2030. But in the long term, I think we'll need GMOs."

Globally, many share Lin's view, including the good people at Monsanto, the world's largest seller of transgenic seeds. In 2009, Monsanto advertisements asked, "9 billion people. A Changing Climate. NOW WHAT?" The answer, the ad said, was "advanced hybrid and biotech seeds . . ." One might expect that from a company with a huge financial stake in the matter—three-quarters of Monsanto's profits came from transgenic seeds in 2007—but Monsanto is hardly alone in promoting GMOs. There is enormous political and philanthropic support for transgenic agriculture, and not just in its traditional stronghold of the United States, where transgenic seeds are used on 90 percent of the soybean acres and in 63 percent of corn production. Proponents of GMOs argue that manipulating genes will enable humanity to boost yields in time to cope with climate change. Worried about future droughts? Let scientists develop drought-tolerant seeds. Don't like the dead soil and polluted water left behind by conventional agriculture? We can engineer seeds that don't need as many pesticides, herbicides, and fertilizer.

It's a tempting vision, and doubtless it will be pursued in some countries. But it is less clear that GMO seeds actually deliver the benefits proponents claim. Aside from unresolved health and safety questions, the commercial production of GMO seeds in the United States thus far does not appear to be substantially better than that of conventionally bred seeds. "Corn has a narrow range of temperatures that the plant will tolerate, and GMOs have not changed those physiological responses," said the USDA's Hatfield. "Just like it would take a lot to change humans' optimal temperature to something different from 98.6°F, we would have to get pretty imaginative to change corn's basic processes with genetic engineering."

Another reason for caution is that GMO technology is controlled by a handful of global corporations whose past behavior has not displayed much public-spiritedness. Monsanto, for example, insists that GMO seeds can benefit the poorest farmers in the world as well as the richest. I heard a Monsanto executive make this claim in a speech to the International Agricultural Management Association, a trade association, in June 2008. According to John Trey Key III, Monsanto's director of seed and trait strategy, Monsanto would double the yield of its corn, cotton, and soybean seeds by 2030. The company would also reduce by one-third the amount of water and other resources the seeds needed. Most remarkably of all, Monsanto's seeds would "be made available to end users royalty-free, whether they're in the United States or Burkina Faso," Key said.

It seemed an extraordinarily generous offer; after all, developing transgenic seeds requires millions of dollars of investment. An audience member asked how Monsanto planned to make a profit on seeds it gave away royalty-free. Key replied that testing seeds in different climates provided valuable clues about how to improve them. That wasn't a terribly convincing answer, so later I asked Key again: How could Monsanto possibly make money selling seeds in Burkina Faso, one of the poorest countries on earth? "We think," he replied, "that if you can create value for even a poor farmer by increasing his yields, you should be able to figure out a way to share that value with him." In other words, if Monsanto's scientists could develop a drought-tolerant seed that increased Yacouba Sawadogo's yields of sorghum by 20 percent, Sawadogo should find it worthwhile to pay Monsanto part of that additional 20 percent. So Monsanto, it seemed, still intended to get paid in Africa; the payment just wouldn't be called a royalty fee.

If Sawadogo and other poor farmers in Africa did give Monsanto's seeds a try, they would find that they had a very different understanding of a farmer's relationship to seeds from that of the company. In the United States, Monsanto has required farmers who buy its seeds to sign contracts that explicitly prohibit them from saving seeds for replanting, and it has gone to great lengths to enforce the prohibition. Investigative journalists Donald Barlett and James Steele have reported that Monsanto has hired private eyes to videotape farmers it suspects of saving seeds. The company has also sued hundreds of farmers and used the threat of lawsuits to intimidate countless more, including organic farmers whose fields were contaminated by transgenic seeds the wind blew from neighbors' fields. Prohibiting the saving of seeds contradicts centuries of agricultural practice the world over; farmers everywhere traditionally put aside a portion of the harvest to plant next year's crop. Monsanto says the prohibition is necessary to protect the company's patents and cover the costs of developing its seeds. Critics say Monsanto—which is the world's largest seed company and enjoys monopoly power in many of the markets where it operates—is exploiting its market power to take unfair advantage: prohibiting the saving of seeds, after all, forces farmers to buy new ones every year.

Needless to say, Key did not mention any of this to the IAMA audience. He spoke instead about how Monsanto was collaborating with the Bill and Melinda Gates Foundation and the Warren Buffett Foundation to create an initiative called Water Efficient Maize for Africa. Under WEMA, Monsanto would share its technology with farmers in South Africa, Kenya, Uganda, and Tanzania. On its website, Monsanto claims that WEMA has put drought-tolerant maize on 400,000 acres of land and achieved 25 to 35 percent increases in yield. Those are very impressive numbers, but there is a catch: the only documentation for them is WEMA's own press release.

Most peer-reviewed research has found little reason for optimism that GMO seeds will revolutionize either production yields or environmental benefits. *Failure to Yield*, a report published in 2009 by the Union of Concerned Scientists, reviewed independent studies of commercial transgenic crops grown in the United States and concluded: "After more than 20 years of research and 13 years of commercialization . . . [genetic engineering] has done little to increase crop yields." The report left open the possibility that transgenics might do better in the future, but it cautioned that improvement was unlikely: apparently the transgenes under consideration for future use produce more complex effects than today's varieties, and early research indicates that some effects could be negative. Defenders of GMOs complained that the UCS study had ignored various benefits of genetically modified seeds. The UCS author, Doug

Gurian-Sherman, replied that the alleged benefits had been observed only in experimental seeds, and since most experimental seeds did not end up getting commercialized, such benefits had little relevance to decision makers.

GMO agriculture took another hit in 2009 with publication of the *Agriculture at a Crossroads* report by the International Assessment of Agricultural Knowledge, Science and Technology for Development. Sponsored by the World Bank, the Food and Agriculture Organization, and other UN agencies, *Agriculture at a Crossroads* was perhaps the most authoritative analysis of agriculture in developing countries ever conducted. A kind of IPCC for agriculture, the Assessment drew on hundreds of scientists from around the world to advise governments on how to reduce hunger, poverty, and disease while still respecting the environment. Testifying before the U.S. Congress, Robert Watson, the director of the assessment, pointed out in the politest way possible that genetically modified crops were an unproven technology whose risks and benefits remained uncertain: "*It is possible* that GM crops could offer a range of benefits over the longer term," Watson said, ". . . however, it is likely to be several years at least before these traits *might* reach *possible* commercial application [my emphasis]." In other words, not only did genetic engineering have no immediate relevance to feeding the world's hungry; it might never have any relevance. African farmers were perfectly capable of raising yields with current crops and inputs, Watson added. What they needed most were political and economic reforms, including an end to northern governments' massive subsidies of their own farmers.

The final drawback to transgenic agriculture is that, like conventional industrial agriculture, it does not appear to be very resilient to the impacts of climate change, especially the volatility in temperatures and precipitation that scientists expect. As climate conditions shift, chances increase that a given seed variety might not be able to adapt. "We absolutely have to develop seeds for improved and climate-adapted varieties, but we also need to increase the diversity of seeds," Sara Scherr said. "[Instead] a lot of the focus is on 'Let's get a few seeds that are drought-resistant that can be used on millions of hectares.' The current business model in agriculture is based on maximizing volume, which militates against diversity. It's not that they're bad guys in agribusiness companies; many of them want to be good guys. But their business models work the other way. So we need to ask how to change the models and give them different incentives."

One of the Few Tricks Still Up Our Sleeve

Some readers may see the last couple of pages as evidence that I am "anti-GMOs," but I assure you I am not. Being the father of a child growing up under global warming has immunized me against ruling out any technology, however unsavory, that might help extricate our civilization from the terrible mess we're in.

When I began working on this book four years ago, my daughter was a gurgling toddler who delighted in sticking her hands into a squishy plateful of mashed potatoes. Now, Chiara is a precocious preschooler who uses words like *problematic* and thinks ahead enough that the other day she telephoned me from a sleepover to remind me to feed the two kittens that had recently joined our household. "And be sure to pet them, Daddy," she instructed. "They like to be petted." As I near the end of this book, I still find it hard to reconcile the joy that is Chiara with the climate disasters that loom before her. The older she gets, the closer those disasters come. The relentless momentum of the climate system assures as much, and the glacial pace of the human response to date only adds to my foreboding. I look at Chiara, at her cheerful countenance, her mischievous eyes, her blond locks, and there is a disconnect. Despite all the research I've done on climate change, I still can't fully take in that this innocent creature, and millions more like her around the world, will have to suffer because grownups insisted on making foolish choices. In my father's heart, I think there must be a way to stop this movie before it gets to what Chiara would call "the scary part." But my journalist's brain knows the truth: at this point, there's no avoiding the scary part; our only hope is to prepare for it as best we can. That includes exploring any and all options that promise to help us.

My environmental friends may not like to hear this, but I'd be more than happy if GMO seeds work as well as their proponents claim. I feel the same about nuclear power and carbon capture-and-storage, as well as many of the ideas for reversing global warming that are classified as geoengineering. These are controversial technologies, and I know that some have very significant drawbacks. However, we no longer have easy choices in the climate fight. My daughter's future, perhaps her very survival, is at stake, so I certainly won't let ideological objections get in the way of accepting help from imperfect sources. If GMO seeds could work as well as Monsanto says, what a blessing! Then it wouldn't matter so much that by the time Chiara is a young woman the Midwest will be enduring scorching summer temperatures. And imagine the benefits for Uma in Bangladesh. Salt-tolerant GMO seeds might enable her coastal village to achieve respectable rice yields again.

The problem I have with GMOs is less the technology itself than the economic interests and political practices behind it. If GMO seeds were shown to work safely, and if Monsanto really were to provide them to African farmers at prices the farmers could afford, who could object to that? But the main reason we hear so much about GMOs as a response to climate change is not that they actually promise to solve the problem; it is that GMOs have rich and powerful interests like Monsanto and the Department of Agriculture promoting them. By contrast, the FMNR methods that have transformed agriculture in the western Sahel are little known, even inside the countries where they are practiced. During a press conference at the Copenhagen climate summit, I had the chance to ask Burkina Faso's president, Blaise Compaore, about FMNR. Though I was careful to sketch FMNR's core principles in my question, it was clear the president had no idea what I was talking about. His reply invoked Burkina Faso's potential for producing biofuel, one more "solution" promoted by special interests that has delivered less in practice than it does in theory.

If public policy were decided purely on the merits, I believe that FMNR would be embraced as a leading agricultural model for the second era of global warming. Not only has FMNR made poor and vulnerable Africans somewhat less poor and vulnerable, but its lessons about improving the soil's fertility and water retentiveness are, as the USDA's Hatfield said, directly applicable to the climate challenges facing farmers in the U.S. Midwest. FMNR offers one more huge potential benefit as well: growing trees could help to reverse the soaring temperatures of global warming.

For years, discussions about climate change mitigation focused solely on reducing ongoing emissions of greenhouse gases. Reducing such emissions is essential, but at this point it is also insufficient. Reducing emissions merely slows the increase in the greenhouse gases in the atmosphere. Temperatures can stop climbing only if the absolute amount of greenhouse gases in the atmosphere declines. That is the goal of many geoengineering schemes: to extract greenhouse gases from the atmosphere or ocean.

But a moment's thought reveals that agriculture offers the most direct means to this end. Through the miracle of photosynthesis, trees and other plants take in CO_2 and let out oxygen. The CO_2 is thereby removed from the atmosphere. Plants then store it—in their leaves, their stalks, their trunks and roots. Eventually, after the plant dies and decomposes, the CO_2 is stored in the soil for some years before it is re-released back into the atmosphere. The earth's plants and soils already hold three times as much carbon as the atmosphere does, Scherr notes: "About 1,600 billion tons of this carbon is in the soil as organic matter and some 540 to 610 billion tons is in living vegetation." Most encouraging, it appears the earth's plants and soils could hold a great deal more CO_2 without upsetting the balance of natural systems.

Although agriculture is key to mitigating global warming, its potential has been largely ignored so far. "Agriculture and forestry is the one sector where we have the technology in hand today to capture and store carbon, yet agriculture is hardly being mentioned in the lead-up to Copenhagen," Scherr said in August 2009. "All the focus is on energy. Billions of dollars are being poured into carbon capture-and-storage for coal plants, a technology that might or might not ultimately succeed. Much less money is being spent on agricultural and forestry techniques that we know work and that also deliver co-benefits, such as poverty reduction, increased biodiversity, and watershed protection."

In the Worldwatch Institute's 2009 *State of the World* report, Scherr identified five strategies that appear especially promising. As it happens, each of them is identical to or congruent with the methods being pursued by Yacouba Sawadogo, Jiang Gaoming, and other practitioners of ecological agriculture the world over. The first imperative Scherr cited is to halt the destruction of forests worldwide and especially in the tropics, where the fast-growing trees are so effective at absorbing carbon. A second option—restoring vegetation in degraded areas—is the essence of FMNR. So is a third: "enriching soil carbon," which includes "no-till agriculture." Instead of industrial agriculture's habit of removing crop residues and plowing up soil before planting—which releases large amounts of CO_2 into the atmosphere—no-till agriculture leaves the residues in place and inserts seeds into the ground with a small drill, leaving the earth basically undisturbed. I saw no-till agriculture being practiced in China's Henan province, and though the farmers had no conception of carbon emissions, they said no-till did produce excellent yields. A calculation by the Rodale Institute, a nonprofit group in New York State, found that if no-till was used on all 3.5 billion acres of the earth's tillable land, it would sequester half of humanity's annual greenhouse gas emissions.

Finally, Scherr notes the potential contribution of biochar. Biochar, which is simply a fancy scientific name for charcoal, is produced when plant matter (e.g., leaves, stalks, trunks, roots), manure, or other organic material is heated in a low-oxygen environment. When biochar is inserted in soil, the effect is to remove CO_2 from the atmosphere and store it underground, where it will not contribute to global warming for hundreds of years. A second effect is to increase the soil's fertility and ability to retain water, which in turn encourages greater crop yields. Normal agricultural and forestry production methods leave behind large quantities of waste materials: corn stalks, rice husks, peanut shells, tree trimmings, manure. If this waste were transformed into biochar and inserted into soil, it would offset 594 million tons of CO_2 equivalent, according to Johannes Lehmann, a professor of agricultural science at Cornell University and one of the world's foremost experts on biochar. If biochar were added to 10 percent of global cropland, Lehmann calculates, it would store 29 billion tons of CO_2 equivalent—roughly equal to humanity's annual greenhouse gas emissions.

Sound too good to be true? George Monbiot, an environmental writer for the *Guardian* newspaper, thought so. Seizing on one advocate's proposal to obtain biochar from vast tree plantations in the tropics, Monbiot blasted widespread deployment of the technology, calling biochar "as misguided as Mao Zedong's Great Leap Backwards." Monbiot was correct that relying on plantations to produce biochar could cause poor farmers to be kicked off their land (to make way for the plantations). It could also cause food prices and hunger to increase (as land that produced food is diverted to biochar), as well as more species loss (as wildlife loses habitat). But Monbiot unfairly tarred all biochar supporters with the same brush, as he himself later admitted. Supporters such as James Hansen of NASA, James Lovelock, the inventor of the Gaia hypothesis, and Johannes Lehmann had clearly stated they did not favor the plantation approach, which applies the mindset of industrial agriculture to a technology that is the essence of ecological agriculture.

To be sure, neither biochar nor FMNR is a silver bullet that will magically solve humanity's food problem in the second era of global warming. But they may be pieces of silver buckshot, to borrow NASA climate scientist Cynthia Rosenzweig's phrase: parts of a solution that could take us a good distance in the right direction. Our very desire for a silver bullet is a manifestation of the industrial mindset that has shaped our approach to agriculture over the past fifty years. For every problem, in this view, there is a technical fix. But this is hubris. Industrial agriculture has brought important achievements but at great cost. More to the point in the peak oil era, it appears to have run its course. It is neither climate-friendly nor climate-resilient, and it is time to phase it out.

Instead of trying to outsmart Nature—to manipulate it with enough chemicals, fossil fuels, and genetic engineering to produce beyond its capacity—we should try to work with Nature. Ecological agriculture is plainly superior for adapting to climate change, and more and more evidence suggests it can also match industrial agriculture's production levels after a brief transition period. This, despite the fact that ecological agriculture has received but a tiny fraction of the R and D monies, government support, and private investment industrial agriculture has long enjoyed. If we were to reverse these political and economic trends, there is good reason to suspect that ecological agriculture is capable of much more.

Meanwhile, each of us can help realize this potential—and enjoy tastier meals and better food security in the bargain—if we begin growing some of our own food. This is a relatively easy step for rural and suburban residents to take, but city dwellers too can be part of the solution. There are empty, unused spaces scattered across many of the world's cities that could be transformed into community gardens. All it takes are a few dedicated citizens, the cooperation of local authorities, and some tools and seeds. Hundreds and perhaps thousands of gardens have sprung up in Detroit alone in recent years after the decline of the U.S.

auto industry led to a massive depopulation of the city. Residents and activists have turned the resulting vacant lots into gardens that are providing residents with fresher vegetables than they've ever enjoyed, even as they create jobs and economic opportunity.

If you think your little garden won't make any difference, think again. There is an inspiring precedent for Michelle Obama's campaign to plant organic gardens across the United States. During World War II, First Lady Eleanor Roosevelt urged Americans to join her in planting Victory Gardens. Families growing their own fruits and vegetables, said Mrs. Roosevelt, would help the war effort by freeing America's professional farmers to concentrate on the corn, wheat, livestock, and other products only they could provide. Mrs. Roosevelt's initiative was opposed by her husband's Department of Agriculture, but she persisted in her advocacy and the American people answered her call. Soon, even city apartments had lettuce, carrots, and other vegetables growing on windowsills. Within two years, Victory Gardens were providing almost half of all the fruits and vegetables being consumed in the United States.

Nature and its processes are more powerful than we know. As we face the second era of global warming, it is clear that our current approach to food is unsustainable. But it is equally clear from the examples of FMNR, biochar, and organic agriculture in general that food also offers a way to turn things around. Agriculture is one of the few tricks humanity still has up its sleeve in the race to avoid the unmanageable and manage the unavoidable of climate change. Let's not squander it.

9 | While the Rich Avert Their Eyes

Climate change is the greatest weapon of mass destruction of our times. Unless we in the rich countries recognize this fact and do something about it, we are guilty of crimes against humanity.

—SALEEMUL HUQ

WATER SCARCITY SEEMS inevitable in California's future, but at least Chiara will be more used to dealing with it than most people. Just before her fourth birthday, our town imposed mandatory water rationing. California was enduring a third consecutive year of drought at the time. Most of the state's residents and businesses had access to water the state had stored in previous years, but our town was not connected to any outside water system. Our drinking water came from a nearby creek whose flow was captured by a small dam. When the winter rains had not arrived by late January, local authorities decided the only recourse was to impose strict limits on consumption. They divided the volume of water remaining behind the dam by the six hundred households in town and came up with a ration of 150 gallons per day per household. That was less than half of the average American household's consumption of 400 gallons a day.

Our town is quite environmentally minded, so there were not many complaints. Although our water consumption was already below the national average, now we had to do more. Our children led the way. At Chiara's preschool and at the local elementary school—"where the big kids go," as she says—the teachers did a great job of explaining the situation in a serious but not scary way. They took the kids on field trips to see the town's dwindling water supply for themselves, and then everyone brainstormed about how to solve the problem—how to use much less water than we were used to using. Most of the suggestions were common sense, but they ended up having a sizable impact on the town's consumption because the kids brought the ideas home and got their parents to adopt them too.

When I asked Chiara what she had done at school one day, she said her class had talked about not wasting water. She then recited the old "If it's yellow, let it mellow" rule for flushing the toilet less often. Other ideas from the kids included: Take shorter showers. Don't let the water run while you wash your hands or brush your teeth. Put buckets under the shower and use the dirty water for toilet flushing. Wash your dishes in a tub. These and other changes reduced the elementary school's water consumption by 35 percent within a matter of days. Thanks to firm guidance from our water authorities—violators of the 150-gallon-a-day limit were given two warnings, and a third violation got their water shut off—the town as a whole was 98 percent compliant. Now it's routine for the people in our town to treat water as a precious resource rather than take it for granted. When a little pal of Chiara's visited from San Francisco and left a faucet running one day, I was proud to see my daughter reach up and turn it off before gently informing her friend, "We don't waste water here."

I still think about whether I should move Chiara to somewhere less vulnerable to climate change. Scientists say that California faces greater impacts than most of the United States, largely because of our state's wide range of topographies and habitats. For example, California has a long coastline, and both its people and its economy depend heavily on proximity to the ocean, so sea level rise and storms are major concerns. Our ports—notably, Oakland and Long Beach—are the entry points for much of America's trade with China and other Asian nations; three feet of sea level rise will put them out of business, unless adaptation measures are taken. Meanwhile, much of our water supply originates hundreds of miles from the people who rely on it, so the water has to be pumped to them—an increasingly risky proposition as global oil supplies peak (California uses electricity, not oil, to pump its water—indeed, water pumping is the single largest use of electricity in the state—but rising oil prices figure to boost the price of competing energy sources as well). Finally, what British science adviser David King called the rising temperature baseline is melting the mountain snowpacks that hold much of California's water. Paradoxically, this melting stands to threaten Californians with too much water at some times, such as spring flood season, and too little at others, especially in summer, when humans, animals, and crops need it the most. Yes, there are many reasons to consider leaving California in the second era of global warming.

But as I consider whether our family should move, I can't help remembering that even being able to ask this question reflects our family's privileged position. We live in one of the richest countries in the world, and we are personally well off, therefore we have options. Most of humanity does not. When I think of Sadia and her family in Bangladesh, I'm confident that her father loves her just as much as I love Chiara. But he has far fewer options for protecting her from climate change, simply because he is a poor man living in a poor country.

Sadia's father is a sharecropper who grows rice, wheat, vegetables, and chiles on land he doesn't own. The day I met him, I remember thinking he looked much older than his actual age of thirty-seven, as poor people often do. He said he had had ten years of schooling—it was one reason he was chosen as mayor of his village—but he was not familiar with the terms *global warming* and *climate change*. All he knew was that if the river continued to wash away his village's land, he and his neighbors would have to leave. They would not have much choice of where to go. Bangladesh is the most densely populated major country in the world; its 156 million people—roughly half the size of the U.S. population—are crammed into an area about as large as Iowa. "We will go to live with relatives," he told me. But the relatives lived only two kilometers away, so they were probably quite vulnerable to flooding themselves.

How are poor people like Sadia's family supposed to adapt to climate change? Adaptation is challenging enough for the comfortable, as this book has described. But for the well off, adaptation is a matter of will, not capacity; if they have the wit to grasp the problem, they have the means to address it. For the poor, even a strong will counts for only so much. Poverty leaves them doubly exposed to climate impacts. It means they tend to live in the places that are the most vulnerable—for example, the lowest-lying parts of a city or country—because they can't afford better. And it often means they cannot take even basic steps to protect themselves, such as choosing better housing.

In Bangladesh's flood-prone capital, Dhaka, rents are cheapest for apartments on the ground floor. Why? Because they flood more often than those above. The west of Dhaka is protected from the Buriganga River by a massive earthen embankment, and one afternoon I hired a taxi to drive me along its crest, a dirt track pocked by deep holes and rocks that must have cracked many an axle over the years. From the crest down to ground level was quite a drop, a distance of at least forty feet. It was late in the day; the sun was setting to our right, beyond the murky flow of the Buriganga. But my eyes were riveted to the left, the area supposedly protected by this embankment. Crowded below me were mildewed building after mildewed building, most of them perched just inches above pools of filthy water. I watched a mother and two small children leave one house and walk along a wooden board to the building next door. A foul odor filled the air, the residue of the many leather and glue factories in the area. I should emphasize that I was visiting Bangladesh in dry season; in the monsoon months, the water level down below would have been far higher. Indeed, in 2004, heavy rains had combined with drainage failures to put 60 percent of Dhaka underwater.

Poverty may be the single most important example of how social context shapes one's capacity for climate change adaptation. In 1991, Bangladesh was hit by a cyclone that made Hurricane Katrina seem almost gentle by comparison. With winds of 140 miles per hour and storm surges of nineteen to thirty-two feet, Cyclone 1991, as it was called, was the equivalent of a Category 5 hurricane. Centered on the southeastern coast of Bangladesh, the storm killed an estimated 138,000 people—approximately one hundred times more than Katrina did. It also demolished countless houses and ruined agriculture throughout the region. At first glance, Bangladesh seems to have been the unfortunate victim of a natural disaster. But the true cause of most of the death and destruction was poverty, argued Muhammad Yunus, the Bangladeshi economist whose championing of micro-lending won him the Nobel Peace Prize. In a study of Cyclone 1991 prepared by the Bangladesh Centre for Advanced Studies (BCAS), Yunus explained, "What makes a disaster is not so much the size of the physical event but the inability of the stricken community to absorb it. . . . When resources are already scarce and the baseline is under pressure, the advent of a disaster stretches the fabric of society to its limit, if not breaking it down."

Atiq Rahman, the director of the BCAS and coeditor of the Cyclone 1991 study, extended the argument by pointing out that the state of Florida was struck by an equally powerful hurricane a year later. Hurricane Andrew was the most destructive hurricane in U.S. history, causing an estimated $26.5 billion in property damage. Its death toll, however, was relatively small: 26 deaths, compared to the 138,000 who perished in Bangladesh. "Had people in Bangladesh been as rich as people in Florida, our casualties would have been much lower," said Rahman. "We would have had better shelters, better evacuation options, and other resources to protect ourselves with."

Subsequent events proved the point. After Cyclone 1991, foreign donors helped the Bangladeshi government to invest in better storm and flood defenses. By 1997, almost two thousand cyclone shelters and two hundred flood shelters had been built. Embankments were constructed along 3,931 kilometers of coastline; 4,774 kilometers of drainage channels were dug. With on-the-ground assistance from NGO activists, an early warning and evacuation system was put in place that enabled the country's weather agency to transmit bulletins to hundreds of remote rural villages, often via cell phone. All this paid off in 2007, when Cyclone Sidr struck Bangladesh. "Some 3,000 people were killed in the Sidr cyclone," said Saleemul Huq. "Nevertheless, I consider it a success story. We evacuated a million people in advance of the storm. Three thousand dead is 3,000 too many, but compare that to 1991, when 138,000 people died."

In the years to come, Chiara's California and Sadia's Bangladesh are projected to face similar physical impacts from climate change, but the economic resources each can bring to bear on the problem are vastly different. California is one of the wealthiest places on earth, Bangladesh one of the poorest. Boosted by the riches of Hollywood and Silicon Valley, California ranked as the ninth-largest economy in the world in 2008. (Measured separately from the rest of the United States, California's GDP is larger than Brazil's and trails only the European Union, the United States, China, Japan, Germany, France, Britain, and Italy.) The state government has had well-publicized budget troubles—during the summer of 2009, it briefly was reduced to issuing IOUs when lawmakers could not agree on a budget—but the larger economy remains strong, vibrant, and deeply capitalized. Google is not going away anytime soon. Meanwhile, half the kids in Bangladesh don't go to school or get enough to eat. About 10 percent of them—between 8 and 10 million children—work full-time jobs, often under very dangerous conditions. "If I don't take home 60 *taka* [$1] a day, someone in my family will go hungry," thirteen-year-old Mijan told the BBC. Eighty-one percent of the Bangladeshi population live on less than $2 a day; 49 percent live on less than $1.25 a day.

With so little to tax, the government in Bangladesh is also poor, and its capacity is further weakened by brazen corruption among the nation's governing elite. California is no stranger to dysfunctional government itself, as the budget follies of 2009 showed, but rich societies can more easily transcend such political foolishness. Even in the midst of the 2009 budget crisis, the state government rolled out an ambitious plan for adapting to climate change—the *California Climate Adaptation Strategy*. By contrast, Bangladesh has yearned for years to implement its own adaptation plans, but it doesn't have the money. Indeed, Bangladesh published one of the world's first climate change adaptation plans, in 2005. In 2008, the government released a more detailed proposal that was as serious and well informed as the plan California released a year later. Most of Bangladesh's plan, however, has remained just that—a plan—because the government cannot find the funds to turn it into reality.

As the second era of global warming advances, watch for this disparity to emerge worldwide: the more forward-thinking parts of the rich world will begin to adapt to climate change, spending large sums of money to protect themselves, while poor countries and communities are left to their own devices. "We must remember that the rich countries are the ones who caused climate change, so we are the ones who should pay to fight it," said Madelene Helmer of the International Committee for the Red Cross. "But that is not happening. The Netherlands, my home country, is spending about $1 billion a year on adaptation, while Bangladesh gets nowhere near this much."

Big numbers like that are often abstractions, but the consequences could not be more concrete for Sadia's village. Despite the threat of future floods, Sadia's father said he would prefer to stay in Antarpara, and the shouted assents of the sari-clad women surrounding us indicated that the mayor's desire was widely shared. "Our hope is that the government will come and build a groin in the river [that is, a concrete barrier to divert flow away from the bank], so we can stay here," Sadia's father said. "Until then, we can do nothing except move our houses when floods wash the land away. I have moved my house four times." Practical Action, one of the many NGOs active in Bangladesh, pleaded the village's case with the Bangladesh Water Development Board, a local staff member told me, and officials agreed the problem was severe. "We urged the government to make protection of this village a higher priority," the staffer told me as our van left Antarpara at dusk, bumping along a crowded one-lane dirt track where boys wearing skullcaps were soliciting donations for the local mosque. "But [the officials] said, 'We have many priorities.' Unfortunately, they are right. The resources of the government are limited, and there are many places like this in Bangladesh."

A "Raw Tension" Between Floods and Droughts

Many places in California are also very susceptible to flooding, starting with the state capital itself, Sacramento. Stein Buer, the executive director of the Sacramento Area Flood Control Agency, told me Sacramento was even less protected than New Orleans had been before Katrina. Sitting in the flatness of the Central Valley, the California capital is straddled by the Sacramento and the American rivers, each of which is fed largely by runoff from the Sierra Nevada. The problem, said Buer, is that the capital is "located at the bottom of a 27,000-square-mile watershed, and all that water has to squeeze by Sacramento on its way to the sea."

At the moment, the only things protecting the people, businesses, infrastructure, and government buildings of Sacramento are levees made of packed earth. Sacramento's levees were built over one hundred years ago, and to a very modest level of safety: after all, they were intended to protect farmland, which could be allowed to flood without great hardship or loss of life. Now these same levees were shielding the government of the ninth-largest economy on earth, along with the 454,000 people who called Sacramento home.

How was this allowed to happen?

"From the time people started to build levees here in the nineteenth century, they looked at the dry land that was created behind a levee and wanted to develop that land," explained Jeffrey Mount, a geology professor at the nearby University of California at Davis. "The development increased the land's value, which increased the financial risk of floods, which led to pressure to build more levees. The new levees created more dry land to develop, and the cycle started over."

Fast-talking with a wrestler's build, Mount at age fifty-two retained a sense of wonder at the nonsense that often passes for conventional wisdom. For example, the 1-in-100-years flood protection Sacramento was said to enjoy was, he said, "actually a very low level of protection. It means that over the course of a typical thirty-year mortgage, the chances of a flood are one in four, because you roll the dice every year." He added, "We've had five one-hundred-year floods in California in my lifetime—in 1951, 1956, 1964, 1986, and 1997."

The 1986 flood came within inches of submerging Sacramento, Mount said as we walked on a pedestrian bridge just across from downtown. The Sacramento River here was about one hundred yards wide. On this sunny day in October, it was flowing calmly, at least fifteen feet below the tops of the levees that extended along both banks. But this stretch of river was "a raging torrent" in February 1986, said Mount, when weeks of heavy rain had swollen it to within three inches of overtopping the levees. "Had the rain continued another twenty minutes, Sacramento's levees would have failed, according to a NOAA study," Mount said.

Fifteen miles up the American River, which joins the Sacramento near downtown, loomed Folsom Dam. Of the approximately 1,400 dams in California, Folsom was widely seen as the most vulnerable to failure. Like the levees around Sacramento, it too was composed mainly of earth, not concrete. Water had surged down the American River during the 1986 rains, but the resulting flooding pales in comparison to what an outright collapse of the dam would do: send roaring downstream a huge pulse of water that could liquefy levees as if they were sandcastles. Climate change increases the risk of such downpours and the catastrophes they can cause. "There are hints that a number of very big dams may be more vulnerable to failures than expected," said Peter Gleick of the Pacific Institute. "The classic example is Glen Canyon Dam [in Arizona]. It wasn't because of climate change, but a flood in 1983 brought that dam very close to failure, to everyone's surprise."

Folsom Dam is slated for an upgrade that should allow it to accommodate more water, but the fundamental problem facing California's dams won't go away, and it's a problem shared by Bangladesh and many other places in the world: climate change will bring both unusually wet and unusually dry years, sometimes one right after the other. That volatility will greatly challenge water system managers in the second era of global warming. "There is a raw tension between the two conflicting demands on dams," said Mount. "Dams are expected, on the one hand, to store water for people downstream to use and, on the other hand, to guard against floods. To prepare for dry years, dam managers will be tempted to let dams fill higher. The problem is, that reduces flood control capacity. If a dam is almost full, and you get a wet year instead of a dry one, the dam doesn't have much room to store the extra water. That's when a dam can overtop or, in the worst case, fail outright."

Meanwhile, more and more people in California are occupying the downstream areas that are at increasing risk from such flooding. The trend toward living in harm's way was rooted, Mount maintained, in Proposition 13, an antitax law passed in 1978 that revolutionized government operations in California. Fueled by a so-called taxpayers' revolt led by landlord and real estate developer Howard Jarvis, Proposition 13 all but prohibited local governments from raising property taxes on existing homes—previously, the localities' major source of revenue. Scrambling to pay for schools, police, and other basic services, local governments began encouraging more development; if existing homes were off-limits to taxes, new homes would have to be built. "Proposition 13 created a grow-or-die syndrome in California," said Mount. "The only way local governments could fund their operations has been to grow—to build more homes, add new developments, occupy more land, even risky land. That has had the effect of putting more people in harm's way."

"It's One Thing to Get Your Feet Wet . . ."

When I asked Mount to name the most vulnerable community in Sacramento, he instantly replied, "The Natomas Basin." Pointing upriver, he said, "You can almost see it there on the right, behind the trees. It's wedged between the American and the Sacramento rivers, just across from downtown. Until ten years ago, that whole area was just farmland. Now there are acres and acres of new houses there, all at extremely high risk."

Mount was something of a maverick within the world of California water policy, but none of the government officials I interviewed disagreed with his judgment about Natomas or the "grow-or-die syndrome" behind it. Maurice Roos, the grand old man of California water experts who served as the state's chief hydrologist, called himself "still skeptical" about climate change when I interviewed him in 2007. He had no doubt about Natomas, though. "I don't think the people who live there understand what risks they face," he said, shaking his head.

At Mount's suggestion, I drove along the top of the levee that shields Natomas from the American and Sacramento rivers. Named the Garden Highway, the levee curls around Natomas for about eight miles. As I drove, I could look down onto the roofs of houses below; I remember seeing a dad pitch Wiffle balls to his son in a side yard. If a flood broke through or over this levee, flooding was inevitable, and not just a little flooding. "It's one thing to get your feet wet," Roos said. "It's another thing to have water over your house." Richard Anderson, the California state climatologist, who happened to join my interview with Roos, brought out a photograph illustrating the threat: an aerial shot of Natomas, altered to reflect the flood levels if the 1986 storms were to recur and the levee were to fail. The water was indeed over the tops of houses. A second shot showed the Arco Arena, Natomas's chief landmark and the home to Sacramento's professional basketball team, the Kings. The arena's parking lot was a lake; the arena itself was half submerged.

But none of this seemed to matter to everyday life in Sacramento. Although these dangers were well understood by government officials and highlighted by the local media, including an extensive series in the *Sacramento Bee* newspaper, the expansion of Natomas rolled on. I visited a large housing development called Four Seasons whose construction was a month from completion. Marketed as "'An Active Adult Community' for people age 55 or better," Four Seasons promised "resort-style living" in "intimately designed homes in a gated, lake-oriented community" where "nature is right at your doorstep." The copywriter might better have left out that last phrase. It's true the river was only half a mile from the development, but the only things standing between the river and the doorsteps of the Four Seasons were flawed ancient levees.

But have no fear, a bubbly saleslady named Claudia told me when I stopped by the rental office: Four Seasons would be a perfectly safe place for my seventy-seven-year-old mother to live if she decided to relocate from back east.

"Are you sure?" I asked. "I read some articles in the *Sacramento Bee* that talked about the flood risks around here. They said the levees aren't so strong." (In the wake of its Katrina humiliation, FEMA had toughened its approach and began decertifying levees across the United States that provided less than 1-in-100-years protection. The Natomas levee was one such levee.)

Her face flushed. "It's very frustrating to see all those stories in the paper," she said, "but I assure you we wouldn't put a planned community in a place we thought was dangerous. The government just wants more money, so it's changing the rules after we did everything we were supposed to. But we are not going away. No way. This is a prime spot, people aren't going to stop coming to the Sacramento Valley, and they want nice homes."

One potential safeguard against river flooding is to create spillways—open areas alongside flood-prone rivers where excess flow can be diverted when necessary. The Yolo Bypass spillway already exists just west of Sacramento. Between three and six miles wide, it doubles as a wildlife preserve; driving or taking the train from San Francisco, you pass through its marshy expanse just before reaching Sacramento. During flood periods, the spillway is capable of diverting 500,000 cubic feet of water per second away from the river's main channel, significantly lowering the risk of downstream damage.

Mount wanted similar spillways created farther up the Sacramento River, and Stein Buer of the Sacramento Area Flood Control Agency agreed. As part of a plan to increase the city's protection to the 1-in-200-years level, said Buer, his agency had proposed paying upstream farmers to allow use of their land as spillways. Such easements would be voluntary, however, which left Mount skeptical that they could overcome the pro-sprawl mentality driving the problem. "Don't think for a moment that we can do it like the Dutch do," Mount said of adaptation in California. "Land use decisions here are made at the local level. If you tell farmers in Sutter County they should become a flood protection zone so Sacramento can live, you know what they'll say? 'You're taking away our development rights.'"

Mount had learned the hard way how risky it could be to threaten development rights in California. The UC Davis professor had been serving on the California State Reclamation Board, which oversees the state's levees, when Hurricane Katrina put most of New Orleans underwater. Interviewed by a local reporter, Mount offered his professional opinion that Sacramento faced a similar risk of devastation. Soon thereafter, the state's governor, Arnold Schwarzenegger, removed Mount and the five other members of the CSRB. Mount told me he was fired because "I was seen as blocking development of new homes, though we never did actually block any." For example, at one point Mount had learned that tens of thousands of homes were being planned in Lathrop, a flood-prone area south of Sacramento plagued by the same poor levees found throughout the valley. Mount and another reclamation board member responded by asking the municipal authorities in Lathrop for an environmental impact statement. The authorities essentially ignored the request, construction proceeded, "and now there are thousands of people living in those homes," said Mount. (A spokesperson from the governor's office denied any wrongdoing, calling the replacement of the board a routine personnel matter.)

"We Need Forests to Be Like Sponges"

Despite his clash with Schwarzenegger, Mount praised the flood control plan the governor released in 2006, which authorized repairs for about 30 percent of the state's shoddy levees, including some of those in Natomas and Sacramento. "There's a long way to go, but it's an improvement over what we had," Mount said. The repairs would cost $4.5 billion over ten years, a sum approved by California's voters—one more example of the advantages of being rich in the face of climate change.

But billions more will be needed to cope with a second threat posed by bad levees—the prospect of losing a substantial portion of Southern California's water supply. The weakest levees in California are found in the Sacramento–San Joaquin Delta, a 1,153-square-mile expanse of land and water that spreads like a fan from San Francisco Bay toward Sacramento. The delta is a hub of California's water supply system, an amalgamation of rivers, pumps, and pipelines that state officials have accurately described as "one of the most complex water storage and transportation systems in the world." The system's basic purpose, however, is simple: it transfers water from the northern third of California, which has 75 percent of the state's supply, to the southern two-thirds, where 80 percent of demand is located. Most of northern California's fresh water begins as rainfall or snowmelt in the Sierra Nevada and ends up in rivers that empty into the delta. From there, two huge pumping stations send the water south through long-distance tunnels, canals, and more pumping stations. The system provides some of the drinking water for 23 million people in Southern California; urban dwellers living as far south as the Mexican border get about 25 percent of their water via the delta, and local sources supply about 50 percent. A much greater amount of delta water is used for irrigation by farms in the San Joaquin Valley, which grows a large fraction of America's fruits and vegetables.

The California water system's greatest point of vulnerability may be the 1,100 miles of levees that crisscross the delta. Originally constructed to create and protect farmland, the levees are now essential to preventing the salty water of San Francisco Bay from contaminating the fresh water the system sends south. The failure of even one levee can compromise these southbound water transfers, as Californians discovered in 2004 with the collapse of a levee around Jones Tract, an island south of Sacramento. The delta's pumping stations were shut down for three days, temporarily halting water extractions intended for Los Angeles and the rest of Southern California.

Earthquakes are the greatest threat to delta levees, but in the second era of global warming, floods and sea level rise will also be problematic. There are numerous fault lines beneath the delta; even a medium-sized quake could quickly crumple the earthen levees above, experts say. Floods, they add, could achieve the same effect, while sea level rise is a more gradual but also irreversible threat. As seas rise, salty bay water will push farther into the delta. This will put increased pressure on the delta's levees, raising the likelihood of their failure, as well as making the water in the delta increasingly salty. Water that is too salty is unfit for drinking or irrigation purposes, so the delta might no longer be able to supply water to Southern California. In that case, some argue, the state might have to replace the delta hub of its water supply system with a man-made structure that would connect the Sacramento River with pumping stations that would convey water south. But there are many environmental, economic, and political objections to the so-called Peripheral Canal—California's voters rejected the idea in the 1980s—and the price tag would be substantial: an estimated $10 billion. A better approach, environmentalists argue, would be for Southern California to stop relying on water transfers from hundreds of miles away. Recalling that Southern California already provides half of its water needs from local sources, these advocates have urged a sharper focus on increasing the efficiency of water use, rainwater harvesting, and reuse of existing water supplies, which happen to be the three top potential sources of "new" water in California over the next several decades, according to the state water plan.

Even Maurice Roos, the California state hydrologist and self-described climate skeptic, worries about sea level rise. "Even 1 foot of sea level rise will make it much harder to save those delta levees," Roos told me, and 1.5 feet would create a situation he called "hopeless." Roos distanced himself from the word *alarming*, which a state document had used to describe the threat, but he added, "It all depends on the amount of sea level rise." Roos was able to be fairly sanguine on this front because he (unwittingly?) put the sunniest possible interpretation on the IPCC's *Fourth Assessment Report*. That report estimated that seas would rise by eight inches to 2 feet by 2100, though as noted earlier, that estimate excluded the role of melting polar ice. "If we get the low end of the IPCC estimate," Roos told me in October 2007, "[sea level rise in the delta] will be a manageable problem." Alas, the *Fourth Assessment Report*'s estimate no longer has much credibility among climate experts.

California's own adaptation plan assumes that sea levels will rise by at least 1 foot by 2050 and by as much as 4.5 feet by 2100.

But again the paradox: along with too much water, California will also have to cope with too little water in the years ahead. The state's adaptation plan does not mince words: "Drought conditions are likely to become more frequent and persistent over the 21st century due to climate change." The main driver of scarcity will be the loss of the Sierra Nevada snowpack, which accounts for roughly a quarter of the state's freshwater supply. Roughly an additional 15 percent comes from the Colorado River, which, as noted earlier in this book, will experience greatly diminished flows in years to come. The rest of the state's water—about 60 percent—comes from rainfall. Rising temperatures will shrink the snowpack by 25 to 40 percent by 2050, according to a California Department of Water Resources (DWR) estimate; if emissions continue to increase along current trends, 90 percent of the snowpack could be gone by 2100. John Andrews, the executive director for climate change at DWR, acknowledged that some scientists consider the 25 percent estimate to be conservative. "But hey," he countered, "25 percent is a big enough number for me. We don't have enough water now to meet total demand."

As a first step, California's climate change adaptation plan called for reducing water use by 20 percent per capita by 2020. The so-called 20 by 2020 Plan would achieve this goal mainly through improving the efficiency of water use. Although Chiara and her schoolmates had helped our town to cut its consumption by nearly twice that much with little difficulty, the state plan angered some Californians enough to provoke lawsuits. State officials, however, held their ground. "Sooner or later, California will be hit by the same kind of prolonged, severe drought that is striking Australia now," Rick Soerhen of DWR told reporter Melinda Burns. "Either we're going to be ready, or the economy takes a terrible hit and people lose a huge investment in landscaping."

The biggest challenge will be getting agriculture to reduce its consumption. The agricultural sector accounts for nearly 80 percent of California's water use, and much of the water is used in remarkably wasteful ways—to grow cotton, alfalfa, and pasture grass, for example, three of the thirstiest crops around. But Big Ag, as it is known in the state capital, wields enormous political power and had beaten back proposed reforms in the past. A study by Gleick and his colleagues at the Pacific Institute found that wise efficiency and conservation measures could reduce agriculture's water consumption by 17 percent—but only if significant changes in regulatory and incentives policy were made—in other words, only if Sacramento stood up to Big Ag.

One of California's leading proponents of smarter water policy was Robert Wilkinson, a professor of environmental studies at the University of California, Santa Barbara, whose squat build and easygoing manner made him a dead ringer for an old buddy of mine. Along with Gleick and Dan Cayan, a professor at UC San Diego who was the chief scientist on the state adaptation plan, Wilkinson was a member of what John Andrews of DWR called "the high priesthood" of California water experts. In the late 1990s, Wilkinson had coordinated the California portion of the U.S. *National Assessment of Climate Change,* a study the Clinton administration commissioned to identify likely climate impacts on various parts of the United States and possible responses. The assessment concluded that water would be one of California's greatest challenges, not least because the outlook was highly uncertain. "You know the joke, don't you?" Wilkinson asked me during one interview "'Under climate change the future is definitely going to be wetter. Or drier. Unless it's both.' That's not an answer people want to hear, because it makes planning much more difficult."

An avid outdoorsman, Wilkinson took every opportunity to inform his scholarly work with visits to the field. I caught up with him and some students one summer while they were passing through Yosemite, the park immortalized by naturalist John Muir. Mesmerized by El Capitan, Half Dome, and the other massive rock formations that loom above Yosemite Valley, Muir helped convince Theodore Roosevelt to make Yosemite one of America's first national parks in 1905. Muir lost a second battle, though. In 1913, Congress passed the Raker Act, which gave San Francisco the right to build a dam across the next valley over from Yosemite Valley. Although Hetch Hetchy Valley remained part of Yosemite National Park, it disappeared beneath three hundred feet of reservoir water after O'Shaughnessy Dam was built. It took twenty years to complete the entire Hetch Hetchy project, including a system of tunnels and pipelines that in 1934 delivered the first drops of water to San Francisco, 160 miles away. Today, Hetch Hetchy continues to supply water to 2.4 million people in the San Francisco Bay Area, while also generating 1.7 billion kilowatt-hours of hydroelectric power per year.

Despite the melting snowpack, San Francisco's water supply was relatively safe for the moment, Wilkinson said as he led his students across the concrete walkway atop O'Shaughnessy Dam. Pointing across the reservoir to the peaks above, Wilkinson explained that the watersheds that feed Hetch Hetchy are higher than their counterparts in most other parts of the Sierra Nevada. The higher elevation means lower temperatures, so the snowpack here would take longer to melt.

But apparently not that much longer. I broke off from the group later to interview Greg Stock, a geologist for the National Park Service who was stationed in Yosemite. Stock was part of a team that had been monitoring Yosemite's iconic glaciers, and he was worried. "The way things are going, Yosemite's glaciers will be entirely gone within a few decades," he told me. "The melting is already having an effect on the visuals available to park visitors. We don't see as many snow-covered peaks as before. But as the melting continues, there will be much more troubling impacts, because glacial melt is the source of the streams and rivers that the flora and fauna of Yosemite rely on. If they dry up, I don't know of any way to replace that water." One likely consequence is more wildfires. Two years later, the state adaptation plan warned that 48 million acres of California, about half of the state's land, "is at a high to extreme level of fire threat. Climate change . . . will further increase the fire hazard."

Around the campfire that night, I asked Wilkinson if he saw any escape from California's water dilemma—the double whammy of stronger floods and deeper droughts threatening at the same time. He surprised me by invoking the old-timers who built California's water system in the early twentieth century: "Chief" O'Shaughnessy, who constructed the Hetch Hetchy dam, and William Mulholland, who masterminded the diversion of water from Owens Valley that enabled the explosive growth of Los Angeles, a feat dramatized in the film *Chinatown*. Environmentalists condemned those acts as crimes committed by greedy scoundrels, but that missed Wilkinson's point.

"Of course many people would say that they were greedy scoundrels," Wilkinson said, "and that what they did was terrible. But you have to admire the boldness of their vision, the sheer audacity of it." When the city fathers of San Francisco sent O'Shaughnessy up to the Sierras to locate a water supply, and he found Hetch Hetchy, it didn't bother him that it was in a national park. In fact, he told his men that they first had to build *another* dam nearby to supply the electricity he would need to build Hetch Hetchy. Then, to pipe the water 160 miles to San Francisco, they built the incredible complex of tunnels and pipelines that still works well today. "It's really remarkable," Wilkinson concluded, "when you think about it.

"What if today's environmentalists had that boldness of vision?" he continued. "We could propose a fifty-year plan for restoring California's ecosystems that would go a long way toward solving our water problems. We've got to build resilience into our ecosystems, especially our forests, if we're going to deal with what climate change will throw at us. Today, our forests are getting drier as more clear-cutting happens, which is one reason we've seen so many wildfires on this trip. When rain hits an area that's been clear-cut, it runs off, whereas if you have healthy forests, the branches slow the rain down so it drips onto the soil and sinks in. Same with snow. If you've got a forest with trees of different ages and sizes, snow can drop through the branches and soak into the earth.

"We need our forests, especially up here in the mountains, to be like sponges," Wilkinson explained. "So when there are big storms or other high-precipitation events, the trees and soil can soak up that water like a sponge. In the short term, that means less flooding downstream. In the long term, it means the water is stored underground and available to use later. Then, if the climate system swerves in the other direction and suddenly there *isn't* enough water coming off the mountains, we can rely on that naturally stored water, both to supply our own needs and to keep the forests less vulnerable to wildfires."

Turning California's forests into water-storing sponges would take a significant investment of time and money, Wilkinson conceded when I asked him about it the next morning. He confessed he had no cost estimate yet. Outlining the fifty-year plan for restoring California's ecosystems in the face of climate change would, he hoped, be the goal of his next phase of scholarly research. "It's time to be bold," Wilkinson told me. "The money can be found if we make a strong case for how to use it wisely. But we have to hurry. To paraphrase [President Obama's science adviser] John Holdren, we're moving from the opportunities to mitigate climate change to the necessity of adapting to it, and the better we do that job, the less suffering there will be."

"Given Half a Chance, Bangladesh Can Manage"

Henry Kissinger had never been to Bangladesh, but that didn't stop him from calling the country a "basket case." The White House national security adviser made the remark in 1971, after the civil war that liberated Bangladesh—then known as East Pakistan—from distant West Pakistan. Perhaps Kissinger was seeking to justify his backing of West Pakistan's army, which massacred tens of thousands of doctors, teachers, journalists, and presumed political opponents on its way to killing more than 300,000 Bangladeshis. In any case, the basket case image has defined Bangladesh in the eyes of the world ever since. "It is still damaging our reputation today," Humayun Kabir, Bangladesh's ambassador to the United States, told me, "but it is not true. We are now feeding ourselves, 140 million people. We have cut our population growth rate in half. All Bangladeshi children are immunized against major childhood diseases. Our economy has grown an average of 5.5 percent a year over the last seventeen years. We are not a basket case at all."

But climate change could well undo all these advances. Bangladesh is the most vulnerable country in the world to cyclones and the sixth-most-vulnerable to floods, according to the UN Development Programme. Two-thirds of the country stands less than sixteen feet above sea level, so sea level rise threatens catastrophe. The three feet of future sea level rise that is now unavoidable will displace an estimated 20 million Bangladeshis. As mentioned earlier in this book, soil and water in coastal regions are already becoming too salty to deliver traditional rice yields. Inevitably, these and other impacts will affect the country's struggle against poverty. "Climate change will severely challenge Bangladesh's ability to achieve the high rates of growth needed to sustain reductions in poverty," said the government's 2008 *Climate Change Strategy and Action Plan*.

To outsiders, Bangladesh may appear doomed in the face of fifty more years of global warming. But spend some time inside the country and things look different. The human factor counts for a lot in adapting to climate change, experts say, and in this realm Bangladesh has great advantages.

"It's usually assumed that poor people will be less able to adapt to climate change, just as it is assumed that people in flood-prone areas are more endangered than those elsewhere," Saleemul Huq told me in London. "But often it is the inexperienced who have more trouble. Look at London: an inch of snow and the transport system collapses, whereas in Edinburgh, where they get snow all the time, life goes on as normal. In Bangladesh, people have been dealing with floods and other disasters for centuries, so they have a greater capacity than rich people who are not used to facing catastrophe. People in Bangladesh have great resilience, and given half a chance, they can manage."

Huq is one of the most influential advocates for the poor within the global climate change debate. Born to a diplomat, he grew up overseas and earned a PhD in biology from Imperial College in London. His empathy with the disadvantaged blossomed after he returned to Bangladesh and cofounded, with Atiq Rahman, the Bangladesh Centre for Advanced Studies, which conducted some of the first studies in the world of how climate change would affect poor people. Huq's first BCAS research project analyzed the nation's fisheries. "I spent months in the river communities of Bangladesh, talking with the fishing families there, who were all quite poor," he recalled. "It was an eye-opening experience for me who had grown up upper-middle class. I got to know the poor as individuals, not as an abstraction. I saw that they were extremely resilient and often ingenious at coping with the circumstances they faced."

His early field research left Huq with an abiding conviction that, as he later put it, "instead of doing research *on* the poor, we should do research *for* the poor. That means going to them, hearing their ideas, and working together to devise and apply remedies." Toward that end, Huq and BCAS developed an approach to climate change adaptation called Community-Based Adaptation. Under CBA, outside experts visit or preferably live in vulnerable communities to hear firsthand what the risks are and how the local people think they should be handled. The outsiders are often the first to explain to locals what climate change is and how it could affect them, but CBA is a dialogue of equals. Experience shows that a top-down approach produces more resentment among locals than political buy-in and progress.

Community-Based Adaptation differs profoundly from what the international community has done to date, Ian Burton, the dean of adaptation scientists, told an adaptation conference Huq and BCAS organized in Dhaka in 2007. "The IPCC and Western governments have focused on what atmospheric concentrations of greenhouse gases will be and what impacts this could produce, which is useful information to have," Burton said. "But they have shown very little recognition of what climate change is doing today to real people living in real places. Now, in addition to the filtering down of knowledge that we've seen so far, we need a flooding up. We have many local case studies of how to adapt to climate change. We need to integrate them into our macro understanding of the problem and generate guidelines local people can use in adaptation."

I saw some of the adaptation measures Burton was referring to when I visited the Bangladeshi countryside. In little green-eyed Uma's village near the southern coast, a forty-year-old farmer named Sausindro had taken a number of simple steps to protect his wife and three kids against the storm surges of cyclones. With NGO help, Sausindro had elevated his mud and thatch house five feet aboveground, establishing it on a mound of packed dirt. Seven hard dirt steps led up to the entrance. Through an interpreter, Sausindro explained that the new structure had served him well: "During the last two cyclones, we did not have to go to the cyclone center."

"Why didn't you go to the cyclone center?" I asked. "Wouldn't it be safer there?"

"The cyclone center is useful," he answered, "but it gets very crowded and there is not much food and water." Having already visited the cyclone center, I found this easy to believe. Built by the German government, the center was located a kilometer away in a neighboring village. The center was locked when I arrived, though this was not as irresponsible as it might sound, for I was visiting in winter, months before the cyclone season. Still, the center was plainly a bare-bones facility, little more than four concrete walls and a roof. "We would rather stay at home if we can," added Sausindro. Then, to show how his family managed this feat, he proudly displayed a *dole*—a basket made from tightly woven strands of bamboo, which, because bamboo floats, would keep rice and fish dry throughout a cyclone's floods.

In a village downstream from Sadia's called Kamarjani, people relied on "floating gardens" that worked on the same principle. Just as their ancestors had done, the villagers wove water hyacinth plants into a watertight mesh; the garden I saw measured about fifteen feet long and ten feet wide and floated in one of the many ponds dotting the village. The mesh was covered with a few inches of topsoil, which was planted with vegetables. Since the structure floated, it was all but immune to flood inundation; high waters simply raised the garden higher.

The same village was also pursuing economic methods of adaptation. The NGO Practical Action had helped to establish a tree nursery that grew mango and other fruit trees as well as medicinal herbs. Seedlings were sold in the local market so other villagers could plant their own trees and herbs, which would eventually provide both direct nutrition and income from future sales. Experts say that diversifying one's income sources in this way is one of the most important ways to adapt to climate change. People who rely on a single income source can easily be wiped out if a flood or drought destroys the basis for their particular livelihood.

After reviewing scores of case studies from around the world, Burton had drawn up guidelines for adaptation in poor communities. His first recommendation echoed the advice of Aalt Leusink in the Netherlands. "We have to adapt now," he said. "We don't need to know how much CO_2 concentrations will increase by 2100 in order to take action today." Burton also disputed the notion that poor communities can't afford to invest in adaptation because they face more urgent problems—shortages of clean water, food, health care, and so forth. "The dichotomy between adaptation and development is false," he said; communities—and countries, for that matter—that did not invest in adaptation would see their economies undermined by climate change, just as adaptation would falter without development to finance it. Burton offered nine guidelines in all, including the argument that poor countries deserve aid from rich industrial nations. But he also cautioned against guidelines, saying, "In the final analysis, adaptation is place-based. It depends on the specific vulnerabilities facing a given place and the strengths and weaknesses people living there bring to bear. Therefore there is no one-size-fits-all approach to adaptation."

Nor was Community-Based Adaptation a sentimental endeavor, Burton emphasized, for it inevitably bumped up against local power relations. How did a given community function—who paid and who benefited? A telling example was shrimp farming, which had emerged in southern Bangladesh as a common means of adapting to sea level rise. One day, driving near the city of Khulna, I saw sweating laborers constructing a shrimp farm. Hacking at oozing gray clay with short-handled hoes, five crews of eight men apiece were pulling forth clumps of soil the size of fists. The clay was then used to construct the walls of the shrimp pond, which was about half as large as a football field. Shrimp farming could be lucrative, which in theory made it "a very good adaptation strategy," said Terry Cannon, a fellow of the Natural Resources Institute of the University of Greenwich. "But in Bangladesh the business of shrimp farming is dominated by about fifty families, while most poor farmers are excluded. So the shift to shrimp farming is actually heightening the conflict between rich and poor in the country."

Social context is also crucial to adaptation in urban settings. It's not difficult to list the changes needed to lower a given city's vulnerability to climate change, even if that city is poor. A paper Huq and a team of coauthors wrote said the essential first step is to remedy existing deficits in infrastructure and services: fix sewers and water pipes, upgrade the public health service, boost emergency response and relief agencies. City governments should also tighten building codes and land use regulations to restrict development in high-risk areas. To make the most progress, city governments should collaborate with community leaders, drawing on local people's knowledge of the risks they face and potential remedies. The problem, Huq and his coauthors lamented, is that most city and national governments in the developing world are too impoverished or too corrupt to turn such proposals into reality. "The need to adapt is being forced onto nations and cities that lack the political and economic basis for adaptation," they wrote, adding, "You cannot adapt infrastructure that is not there."

"This Is Not Charity, This Is Compensation"

Despite its frequently dysfunctional government, Bangladesh has done more over the past twenty years to understand and adapt to climate change than any other country in the world except for Great Britain and the Netherlands. With help from foreign donors, the government has invested $10 billion (in 2007 dollars) to bolster the nation's defenses against floods, cyclones, and droughts. It has also pursued climate-focused agricultural research and made apparent breakthroughs. In 2009, scientists at the Bangladesh Rice Research Institute announced they were in the final stage of testing three varieties of conventionally bred rice that could survive immersion in salt water for longer than two weeks. If that proved to be true, it could help farmers cope with flash floods that mix sea and river water, a scenario likely to occur more frequently as sea levels rise.

Nevertheless, the intensification of climate change in the years ahead means that much more must be done.

Because Bangladesh began studying the problem so early (thanks to Huq and his colleagues at the BCAS), it has a good understanding of the measures needed to climate-proof the country. The government's action plan rests on "six pillars," including food security, infrastructure improvement, and "capacity building"—the last, policy-wonk term refers to hiring and training people capable of carrying out the policies in question. Thus the capacity-building passage says "climate change cells" should be established in every ministry of the government; the technocrats hired for these jobs will "revise . . . all government policies (sector by sector) to ensure that they take full account of climate change." That will require lots of education; few of the country's current technocrats know much about climate change. The infrastructure section calls for repairing existing embankments and drainage systems (the embankment in western Dhaka comes to mind) and building new ones where necessary. It also urges the delivery of fresh water and sanitation services to coastal areas (like green-eyed Uma's in southern Bangladesh) where sea level rise has contaminated drinking supplies. Not surprisingly, the plan explicitly endorses the Community-Based Adaptation approach championed by Huq and BCAS, calling it crucial to boosting the resilience of vulnerable social groups, especially women and children. Huq praised the plan's call for broad public education on climate change, including through the media. "We will be living with climate change for the rest of our lifetimes and our children's lifetimes," he told me. "We must increase adaptation awareness and capacity in all parts of society, so that everyone is working on how to deal with this problem."

The outstanding question, of course, is where the money to pay for all this will come from. Precise cost calculations had not been made as of press time for this book, but the action plan estimated that the first five years of work would cost about $5 billion. Bangladesh knows it must pay some of this bill itself. But it also knows it cannot carry the whole load itself and, further, that the rich industrialized societies that caused global warming are legally obligated to provide most of the funding. The action plan phrases the issue this way: "Adaptation to climate change will place a massive burden on Bangladesh's development budget and international support will be essential to help us rise to the challenge. . . . We call on the international community to provide the resources needed to meet the additional costs of building climate resilience."

The legal obligation of the rich industrial countries is spelled out in the United Nations Framework Convention on Climate Change, the treaty that nearly all the world's governments (including the United States) signed at the Earth Summit in 1992. The treaty, which remains in force today, stipulates that so-called Annex 1 countries—that is, the industrialized nations, including the United States, Japan, Canada, Australia, Russia, and much of Europe—are obliged to help the world's forty-eight least developed countries adapt to impacts of climate change. The reasoning behind this obligation is straightforward. "It is poor countries that are suffering the brunt of climate change," Huq explained, "but it is the rich countries' greenhouse gas emissions that caused this problem in the first place. If we follow the principle of 'the polluter pays,' they are obligated to pay damages. It is important to understand that this is not charity, like the money given to poor countries for economic development. This is compensation."

The climate compensation argument annoys some in the rich industrial world; at the Copenhagen climate summit, President Obama's chief climate negotiator, Todd Stern, said he "absolutely" rejected the suggestion that the United States owes a "climate debt" to the rest of the world. Stern may have been speaking with congressional Republicans in mind, but it is also true that for many years the rich were unaware of the damage they were doing to the earth's climate. Who knew back in the 1960s that driving cars and running air conditioners would lead to such terrible consequences? But as a lawyer himself, Stern surely knew that in a court of law damages are damages, regardless of one's intent. For example, imagine that your grandfather, grown forgetful in old age, turned on the bathtub faucet in your apartment one day and forgot about it until the tub overflowed. The water then soaked through the ceiling of your downstairs neighbor and ruined her furniture. The fact that Grandpa didn't mean to make a mess would be no excuse; you would still have to replace your neighbor's furniture.

One challenge for the compensation argument is that no one knows exactly how much adaptation will cost—not in the rich industrial countries and certainly not in the poorest countries. For years, specialists cited a figure published by the World Bank— $40 billion a year in poor countries—even as they cautioned that this was only a preliminary estimate. Since then, numerous analyses have been undertaken. Perhaps the most authoritative came from Martin Parry, the British geographer and former co-chair of the IPCC, who led a team of scientists who examined the question in detail and published their results four months before the Copenhagen climate summit. Parry and his colleagues found that the World Bank's $40-billion-a-year calculation grossly underestimated the true cost of adaptation for developing nations. Indeed, even a lesser-known calculation by the IPCC, which put the cost at between $40 billion and $170 billion a year, fell well short. The real costs were likely to be two to three times higher. The UNFCCC estimate, explained Parry, ignored many adaptation imperatives. Incredibly enough, in the water sector it excluded the costs of adapting to floods. In health, it considered only the costs of coping with malaria, diarrhea, and malnutrition—obviously important illnesses, but accounting for less than half of the total anticipated disease burden. A number of critical sectors, including tourism—essential to many poor nations' economies—were ignored entirely. Parry emphasized that his own study was not definitive; it only underlined the need for more detailed research. But the scale of the problem was clearly much greater than assumed.

In their rhetoric, rich countries have accepted their obligation to help developing nations adapt to climate change, but in practice they have mocked it. In 2001, a special fund was created within the UNFCCC to fund adaptation by developing countries. Rich nations insisted that donations to this "LDC Fund" (LDC signifies less-developed countries) be voluntary, which doubtless helps explain why so little money has been donated. Since 2001, rich countries have pledged to provide $18 billion but have actually disbursed less than $1 billion, according to an analysis of official data conducted by the *Guardian* and confirmed by the UN. What's more, wrote the *Guardian*'s John Vidal, "Most of the money promised for climate change comes out of official aid budgets, leaving less for health, education and poverty action."

The same potential loophole appeared to undercut U.S. secretary of state Hillary Clinton's proposal at the Copenhagen summit for rich nations to provide $100 billion a year by 2020 for climate action in poor countries. Clinton's proposal, which covered funding for both mitigation and adaptation, was endorsed in the so-called Copenhagen Accord, the side deal arranged by the world's largest greenhouse gas emitters that did not receive formal backing from the summit as a whole. Also included was a short-term target of providing developing nations $30 billion over the three years beginning with 2010. No doubt, $100 billion a year is a serious amount of money, as Mrs. Clinton pointed out in her announcement in Copenhagen. Still, Clinton described the $100 billion as a goal, not a commitment, which raises questions about it materializing. And, of course, even the $100 billion figure falls woefully short of the amount actually needed. Nor is it clear where the money will come from. The text of the side deal refers to both public and private funds, as well as "alternative sources of finance." This phrasing suggests that governments hope to persuade investors to join them in assisting the poor, presumably through money raised from cap-and-trade and other forms of carbon markets. Good luck with that; the record on carbon markets so far is not terribly encouraging.

Under such circumstances, there is no way Bangladesh or any other poor country can implement effective adaptation. The irony is that helping poor nations adapt to climate change is in rich nations' own interest, in more ways than one. Great Britain is a partial exception to the tendency of rich industrial countries to stonewall their adaptation obligations, precisely because it sees the advantages such funding offers to its own adaptation agenda. In Bangladesh, Britain has been subsidizing a substantial program to raise roads, wells, and houses above the level of the last major flood. "Bangladesh is a showcase of what will happen under climate change," Penny Davies, a diplomat at the British High Commission in Dhaka, told me. "It amounts to a testing ground for what island states, including Britain, will need to do to protect ourselves in the years ahead."

The security issues ventilated earlier in this book also argue in favor of helping poor nations with adaptation. One month after Barack Obama was elected president, the U.S. military's National Defense University held what amounted to a replay of Podesta's war game, with Bangladesh again a focus of attention. This time it was severe flooding that supposedly sent hundreds of thousands of Bangladeshis streaming across the border with India, sparking conflict and presenting the outside world with a humanitarian crisis. Such a crisis is in no one's geopolitical interest. Wouldn't it make more sense if the United States and other countries instead worked with Bangladesh and India in advance to increase their resilience so their people don't have to flee if disaster strikes? As Eileen Claussen, a former Clinton administration official who led the U.S. delegation in Podesta's war game, diplomatically put it when offering adaptation aid to the Indian delegation, "We want to help you stay where you are."

If rich nations do not provide climate change aid, said Saleemul Huq, they might instead become targets of climate change terrorism. By no means did Huq advocate terrorism; he was simply speculating that it could emerge if the poor continue to feel mistreated. "[Providing money] would convey that the industrialized nations recognize that the problem exists, acknowledge their historic responsibility for it, and want to address it," Huq told me. "On the other hand, refusing to provide money sends a message of 'We don't care.' As we have seen with terrorism in other contexts, it is the perception of injustice that matters as much as any actual injustice. In ten years' time, you could see terrorists blowing up SUVs in Houston. If Western politicians don't realize this, they will be taking us down a very dangerous path."

Al Gore, the archbishop of Canterbury, and many other political and religious leaders have argued that climate change is at bottom a moral issue, and nothing illustrates the point better than how differently the world's rich and poor are connected to the problem. At the moment, the blameless of the world are suffering first and worst from climate change, while the rich avert their eyes and guard their bank accounts. The poor do generally have greater personal resilience, which will serve them well, but in the end there is no substitute for money. If all the world's people are to avoid the unmanageable and manage the unavoidable in the battle against climate change, their societies must have the financial resources needed to do the job. For rich industrialized societies to continue to withhold the adaptation money that is owed to Bangladesh and other poor nations is legally dubious, strategically foolish, and ethically shameful—a sentence of misery unto death for Sadia, her father, and hundreds of millions of others around the world whose only crime is being poor.

10 "This Was a Crime"

That's what hope is: imagining, and then fighting for, and then working for, struggling for, what did not seem possible before.

—BARACK OBAMA, forty-fourth president of the United States

THEY SAY THAT EVERYONE who finally "gets it" about climate change has an "Oh, shit" moment—an instant when the pieces all fall into place, the full implications of the science at last become clear, and you are left staring in horror at the monstrous situation humanity has created for itself. In twenty years of covering climate change, I've had my share of such moments, but three stand out.

My first came in China in January 1997, after I spent six weeks traveling throughout the country witnessing its feverish economic development and the appalling environmental consequences. Everywhere, it seemed, the land had been scalped, the water poisoned, the air blackened with coal smoke. Yet who could tell the Chinese to stop? My interpreter, Zhenbing, had grown up in a typical Chinese village about two hundred miles northwest of Beijing. Many people there, including his family, were so poor that they could not afford to buy coal (even though they lived near some of the largest coal reserves on earth). As a result, the inside walls of their houses were frosty—or, as Zhenbing put it, "white with icy water drops"—throughout the winter. In fact, Zhenbing told me, when a girl in his village was preparing to marry, her parents would check to see if her suitor's family had white walls or not. If not, it meant the suitor was prosperous enough to be a good catch. The Chinese government and people were understandably determined to transcend such poverty, and cheap coal was their ticket out. Combine this social context with China's gargantuan population, and I could practically smell the massive pulse of carbon that was about to be unleashed on the world.

My second "Oh, shit" moment was described at the beginning of this book. It happened in London, in October 2005, on the day David King made me realize that the global warming dilemma was worse than I'd thought. Dangerous climate change had begun one hundred years sooner than scientists expected, and it was guaranteed to get worse before it got better, locking my infant daughter into a perilous future.

My third "Oh, shit" moment came in New Mexico in July 2009, when Hans Joachim Schellnhuber showed me how difficult it was going to be to keep climate change within survivable limits. Schellnhuber was the chief climate adviser to the government of Germany, the dominant power within the European Union and the world's fourth-largest economy, during the time Germany held the presidencies of the EU and the Group of Eight. He and I had gotten acquainted by telephone during my *Vanity Fair* reporting in 2005, when he had given me a number of extensive, enlightening interviews. When we met in person four years later at a conference at the Santa Fe Institute, he didn't look at all like what I'd imagined. His telephone voice had been full and forceful, the essence of German competence. In the flesh, he was pale, slender, balding. His most striking feature was his eyes, which radiated a relentless, almost unnerving intelligence.

Schellnhuber is a physicist whose specialty, fittingly enough, is chaos theory; some of his most important scientific papers outlined the potential of global warming to push the climate system past "tipping points" and trigger nonlinear effects, such as the shutdown of the Gulf Stream currents that make northern Europe a hospitable place to live. German chancellor Angela Merkel is also a physicist by training, which perhaps accounts for the close relationship she and Schellnhuber were said to have. The week before the Santa Fe conference, with Schellnhuber's support, Merkel had helped convince the other leaders of the

Group of Eight rich industrial nations to aim to limit global warming to 2°C above preindustrial levels.

"Probably [the G8 leaders] agreed to this because they didn't know what it would mean," Schellnhuber joked as he began his keynote speech at the Santa Fe conference. A study Schellnhuber and his colleagues at the German Advisory Council on Global Change—the government's top climate body, known by its German acronym, WBGU—had just examined the question in detail. The study was so fresh, Schellnhuber had not yet briefed Merkel on the results, which were nothing short of breathtaking.

Schellnhuber began by noting that in five months' time the world's governments would gather in Copenhagen to try to agree to a treaty to replace the Kyoto Protocol, which was due to expire in 2012. The Copenhagen meeting was widely seen as humanity's last good chance to avert catastrophic climate change; the WBGU study was intended to find a scientific basis for reaching an agreement all sides could live with. Toward that end, Schellnhuber reminded his audience that average global temperatures had risen by approximately 0.8°C over the past century as the atmospheric concentration of carbon dioxide increased to 390 parts per million. The momentum of the climate system meant that those 390 ppm would also cause an additional 0.6°C of temperature rise, even if emissions magically halted overnight. Keeping the eventual increase within 2°C would therefore require bringing down atmospheric concentrations fast. Schellnhuber did not favor relying on geoengineering to solve this problem; like many scientists, he believed it would "only make things worse." Right or wrong, that assumption meant that global greenhouse gas emissions had to fall at incredible speed.

To have a two-out-of-three chance of meeting the 2°C target—"worse odds than Russian Roulette," Schellnhuber wryly observed—the world's leading economies had to decarbonize completely within ten to twenty years, according to the WBGU study. In other words, they had to reduce their greenhouse gas emissions by 100 percent within the next two decades. Specifically, the United States had to cut its emissions by *100 percent* by 2020—in other words, quit carbon entirely within ten years. Germany and other industrial nations had to do the same by 2025 to 2030. China had only until 2035, and the world as a whole had to be carbon-free by 2050.

Now, Schellnhuber knew perfectly well that expecting the United States to quit carbon by 2020 was unrealistic. Nor did he think Germany, despite its relatively green policies, could quit carbon by 2030, or that China could quit by 2035. But he and his colleagues had identified a possible way around this problem: emissions trading. If big polluters like the United States, Germany, and China could "buy" the right to emit greenhouse gases from poor and other low-emissions countries, the study estimated, the big polluters could delay their deadlines for quitting carbon by ten years or more.

But even the extended deadlines were "brutal," Schellnhuber said, and much of the reason was a crucial political assumption underlying the WBGU study, which was later published under the title *Solving the Climate Dilemma: The Budget Approach.* The assumption was that the right to emit greenhouse gases is shared equally by all people on earth. Known in diplomatic circles as the per-capita principle, this approach was embraced in the WBGU study for reasons of both morality and diplomacy. Developing countries had long insisted on the per-capita principle, arguing that they had as much right to burn fossil fuels as industrialized countries did, especially since the industrialized countries had gotten rich in the process, while developing countries were still poor. Thus endorsement of the per-capita principle was seen as essential to reaching a global agreement that included China, India, and other emerging economies where tens of millions of people were still living in poverty. The problem was, following the per-capita principle meant that big historic polluters like the United States had already used up most of their carbon budget. Humanity's total budget, according to the WBGU study, was 750 gigatons of additional CO_2 emissions through 2050. Limiting global emissions to that amount would give humanity the two-out-of-three chance to honor the 2°C limit that Schellnhuber quipped was worse than Russian Roulette. Divide those 750 gigatons by the forty years ahead and the 2010 global population of 6.9 billion people and it resulted in an annual quota of 2.8 tons of emissions per person.

That was welcome news if you lived in Burkina Faso, Schellnhuber said. It meant your per-capita emissions could actually increase slightly, so someday you might be able to buy a motorbike, perhaps even a refrigerator. Or you could choose to sell your unused emissions rights on the world market. Global demand figured to be strong, for the 2.8 quota was harsh on the biggest historical emitters. Americans had one of the highest current levels, with annual per-capita emissions averaging 20 tons per person. That explained why the U.S. deadline for quitting carbon was the most imminent, but other industrial countries were not far behind. Germany was much more energy efficient than the United States, Schellnhuber said, yet Germany still had to reduce its emissions by 100 percent by 2030. Other EU nations had dates even earlier, in the mid-2020s.

The big surprise, though, was China. "People always think that China will benefit from the per-capita principle," Schellnhuber said. Indeed, I recalled Chinese officials lecturing me during my 1996–97 visit about the sanctity of the per-capita principle. "What do you expect us to do?" one asked rhetorically. "Go back to no heat in the winter?" But Schellnhuber explained that because China had already burned massive amounts of coal and had so many people, it had to decarbonize by 2035.

Back in 2007, Chancellor Merkel surprised almost everyone when she endorsed the per-capita principle at the annual summit of the G8 nations, the only G8 leader to do so. Schellnhuber told me one reason Merkel had done so was that she was the daughter of a Protestant minister and fairness mattered to her. But now that the WBGU study had calculated the real-world implications of this moral stand, even Schellnhuber doubted that the chancellor would renew her pledge—the numbers were just too dire. "I myself was terrified when I saw these numbers," the German climate adviser told me over a glass of wine later. He said the deadlines could be relaxed somewhat if the world settled for merely a fifty-fifty chance of hitting the 2°C target, "but what kind of a precautionary principle is that?" he asked. "We may as well flip a coin." Likewise, he would much prefer having a three-in-four rather than two-in-three chance of hitting 2°C, but that would give humanity even less time to decarbonize. And yes, one could jettison the per-capita principle—as a political matter, this was the path of least resistance for Western leaders—and thereby gain industrial countries another decade or two to quit carbon. But how then to convince China, India, and other emerging economies to limit their emissions? And without such limits, temperatures were certain to rise well past the 2°C target.

As if all this was not troubling enough, Schellnhuber then reiterated a point he'd made in his remarks to the conference: the 2°C threshold was itself by no means as safe as some people supposed. "It is a very bad compromise," he said. "I was instrumental ten years ago in establishing the 2°C target, first within the German government and later within the EU. At the time I thought it was a decent guardrail. Now, we know it's quite dangerous." Nevertheless, he added, "We can't do much better. Even a 2°C target requires us to reinvent industrial society."

So the task of avoiding the unmanageable and managing the unavoidable had been put in a harsh new light. Most of this book has concentrated on the second half of that task: coping with the inevitable impacts of climate change. But the WBGU study is a bitter reminder that we can't live through climate change if we don't also keep global warming within manageable limits, and this, it appears, will be much harder than expected.

It's tempting to hope that Schellnhuber and his colleagues got it wrong, which is always a possibility in science. Their findings are, however, consistent with a growing number of other analyses. NASA's Hansen and other eminent scientists calculated in 2008 that 350 parts per million of carbon dioxide was the most the atmosphere could hold and still maintain a livable climate for civilization. A few weeks after the Santa Fe conference, the UK government's Meteorological Office warned that business-as-usual global emissions growth would lead to 4°C of temperature rise by 2070. The Met Office study didn't explicitly say so, but a 4°C temperature rise would create planetary conditions all but certain to end civilization as we know it and leave many millions of people dead. To cite one example among many, mountain snowpacks would be doomed, crashing water supplies in many places, including California. Worse, positive feedbacks could make runaway global warming all but inevitable, for 4°C would cause the tundra to thaw and the Amazon to burn, releasing vast additional amounts of carbon dioxide.

Some have argued that Hansen and Schellnhuber do the climate fight no favors by publicizing such gloomy projections: laying out such demanding deadlines, the critics warn, only convinces people that the task is impossible. Schellnhuber rejected this critique. "As scientists, we have to tell the bleak, brutal truth," he said during the question-and-answer period in Santa Fe. Invoking the metaphor of traveling on an ocean liner, he added, "The data tell us we are approaching an iceberg. It's no more than a mile away. Well, you have to tell the captain that. It's his decision what to do about it. Maybe he decides the ship is strong enough to handle it. Maybe he thinks we must change course. But as scientists, we must be fearless in saying what the situation is. Then it is up to the public and their leaders to decide what to do."

If Schellnhuber and his colleagues are right, we face a towering challenge. Countries that today are all but addicted to fossil fuels must quit carbon within the next two to three decades. Deforestation and other climate-damaging activities must also be brought to a halt worldwide. And even poor and emerging economies must halt almost all emissions by 2050. Yet even if we manage all this, it will give us merely a two-out-of-three chance to limit temperature rise to 2°C above preindustrial levels, itself an achievement of dubious merit, for it will mean the loss of most of the world's coral reefs, the disappearance of most of its mountain snowpacks, and enough sea level rise, eventually, to inundate the existing coastlines on every continent.

Oh, shit.

"This Was a Crime"

Like me, Schellnhuber was the father of a young child. This fact, divulged as he sipped the last of his Sauvignon Blanc, came as a surprise, for he looked to be in his sixties. It turned out that his first wife had died, and after he remarried, he was blessed by the birth of a son. Zoltan, a name chosen in honor of Schellnhuber's mother's Hungarian homeland, was "a sweet little boy," his father said, who was now almost one year old. Thus the German scientist was almost as new a father in 2009 as I was in 2005 when I had my "Oh, shit" moment with David King in London.

Schellnhuber had invoked both children and grandchildren earlier that day in response to a challenge from *New York Times* climate reporter Andrew Revkin. The scientist had been arguing that there was still hope for humanity, despite the findings he had outlined. To wit, he wanted governments to agree at the Copenhagen climate summit to launch what he called a "Green Apollo" project. Like U.S. president John F. Kennedy's pledge in 1961 to land a man on the moon within ten years, which NASA's Apollo project duly accomplished, a Green Apollo project would aim to shift the world's major economies to low-carbon technologies within ten years' time. Schellnhuber said Germany already had put in place a package of measures that would reduce its emissions by 40 percent by 2020, but much more needed to be done, especially in the United States. "We have the technologies needed to decarbonize our societies," he told the conference, citing improved energy efficiency, thermal solar power, a smart grid, and others. But governments had to provide leadership, in particular by shifting incentive structures and market regulations to send a price signal that would drive private capital and consumers to respond accordingly. "It will be very difficult, but technically it can be done," he added. "The laws of nature are not against us, but they will be if we wait another ten years."

Revkin spoke up to say, "Maybe the laws of physical nature aren't against us, but the laws of human nature seem to be."

Schellnhuber held his ground, however. "Just as there are many laws of physics, there are many laws of human nature," he replied. "On the one hand, there is selfishness and short-term thinking. On the other hand, there is compassion and concern for one's children and grandchildren. We'll see which one wins."

Nevertheless, Schellnhuber told me later that he felt "very sad and very angry" when he contemplated the future his infant son had to look forward to. "It breaks my heart even to think about it," he said. "By 2080, Zoltan will be in his seventies, and he could lead a very miserable end of his life. It must be very unpleasant to be old and fragile without a functioning society around you. Today, if I have a medical problem, it is taken care of. But if our generation does not reinvent industrial society, then my son, and your daughter as well, will have a terrible end of life."

Even a 2°C temperature rise will bring great difficulties for our children's generation, Schellnhuber reiterated, especially for those born into vulnerable circumstances. "I think Zoltan could be fine in a 2°C world, at least for much of his life, because I'm in a privileged situation. I live in a rich country, I am personally well off, and I will do all I can to leave him well prepared. But I know that many other children around the world will not be in a good situation. And that is very vexing to me, very vexing.

"This was a crime," the German scientist added, his voice rising. "It is the result of the lost decade under George W. Bush, the crime of not taking action these past ten years." Global greenhouse gas emissions had accelerated at an unprecedented rate during the Bush era, increasing by 3 percent a year. As a result, said Schellnhuber, humanity had spent 30 percent of its twenty-first-century carbon budget in the first ten years of the century, leaving less time to complete the transition to zero-carbon alternatives. As the world's richest nation and leading source of cumulative emissions, the United States had an obligation to lead the fight against global warming, Schellnhuber said. Instead, Bush had refused to limit emissions at home while also discouraging action abroad, thereby leaving the door open for China and other nations to continue increasing their own emissions. Bush, however, was not the only American Schellnhuber blamed. He was also unhappy with America's voters; after all, they had elected Bush not just once but twice. "What I really don't understand, the real crime," he said, "was reelecting him. How could that happen?"

The First Lost Decade

Only after Schellnhuber left and I was reviewing my notes on our conversation did I realize that his reaction was virtually identical to the one I had had on Westminster Bridge the day I met David King. Facing the fact that my infant daughter was doomed to grow up under ever worsening climate change, I had felt sad, angry, and convinced that a crime had been committed.

Four years have passed since that moment on the bridge, and I have to say that *sad* now seems too small a word to describe how I feel. This book is almost finished, and in the process of researching it I have learned a great many encouraging things about what can be done to cope with climate change. But damn it, the science keeps getting darker and darker. It is now very hard to see how we can avoid at least 2°C of temperature rise, which itself will be extremely challenging to cope with, and we may well encounter a considerably hotter future than that. When I think about Chiara confronting a world like the one described in the UK Met Office's study of a 4°C temperature rise—snowpacks gone, the Amazon burning, sea levels soaring—well, I can't think about it for long. It's too depressing. I said earlier in this book that denial isn't much of a survival strategy, and I still believe it, but I confess I sometimes see its attractions.

But wait a minute, I tell myself. My brothers and sisters and I grew up in the shadow of the atomic bomb, the twentieth century's ultimate nightmare, and we lived through it. I'm not old enough to remember the Cuban Missile Crisis of 1962, but as a young reporter in Washington in the 1980s I had a front-row seat for the belligerent jousting between the nuclear superpowers, when the massive arsenals of the United States and the Soviet Union were poised on hair triggers. I wrote a lot about the arms race in the 1980s, and there were times the facts left me pretty depressed. Yet humanity ended up dodging the nuclear bullet, at least for the time being, and it did so thanks to what at the time seemed rather unlikely developments. Who would have guessed that a radical reformer like Mikhail Gorbachev would somehow rise to the top of the repressive Soviet system and make peace with Ronald Reagan, a right-wing zealot who never met a weapons system he didn't like? It's a useful reminder: history is full of surprises, and sometimes it really is darkest just before the dawn.

But this line of thinking brings only so much consolation, I'm sorry to say, for there is a fundamental difference between the climate crisis and the nuclear arms race. The difference centers on timing. With nuclear weapons, as long as neither superpower pushed the launch button, averting disaster remained plausible. Each side could build thousands of super-weapons and brandish them as aggressively as it wished, but as long as none of the weapons actually exploded, humanity could still back away from the abyss: the two superpowers could choose sanity, take their fingers off the triggers, and start dismantling their warheads, as indeed they began doing in the 1990s. Of course, the nuclear danger remains today; each side still possesses enough firepower to bring an end to modern civilization, and other states and terrorist groups are trying to acquire their own nuclear weapons. Still, humanity is clearly safer from nuclear self-destruction today than it was when I was growing up in the 1960s and 1970s.

Unfortunately, the climate threat cannot be defused so quickly. "This problem will be much harder to solve, because you can't just have two men sit down at a table and agree to stop being stupid," Hubert Reeves, the head of France's national scientific research center, said when I interviewed him in 1991 for *Earth Odyssey*. Reeves was referring to the fact that reversing global warming would require sharp reductions in humanity's consumption of oil, coal, and other fossil fuels—the very life-blood of modern society. In the years since I spoke with Reeves, the task has proven as difficult as forecast; global greenhouse gas emissions rose by 1.1 percent a year during the 1990s, then surged to a 3 percent annual growth rate between 2000 and 2007.

And even that isn't the worst of it. I actually find it relatively easy to imagine humanity shifting to greener sources of energy. After all, we have most of the technologies in hand, many big investors see an opportunity to profit handsomely from deploying them, and scientific necessity leaves little other choice. More troubling is the second reason why the climate crisis is harder to solve: the physical inertia of the climate system. We could have two or two hundred men sit down and agree to stop being stupid—they could even agree to mount the Green Apollo project Schellnhuber urges—and our civilization would still be locked in to worsening climate change for many years to come. Thanks to our past decades of delay, average global temperatures are all but certain to reach 2°C above preindustrial levels, probably in the lifetime of Chiara and Zoltan. And we now know that 2°C is by no means a safe level.

All of which reminds me why, that day on the Westminster Bridge, I felt not only sad and angry but convinced that a crime had been committed. By then, I had spent five years watching Bush and especially Dick Cheney (arguably the real president during the Bush years, as revealed in journalist Barton Gellman's book *Angler*) do all in their power to thwart action against global warming. Among hundreds of examples, the Bush-Cheney administration reneged on Bush's campaign promise to regulate carbon dioxide as a pollutant; it repudiated the Kyoto Protocol; it urged the IPCC, at the written request of ExxonMobil, to fire the chairman of the IPCC; it installed a former lobbyist for the American Petroleum Institute in the White House Council on Environmental Quality, where he censored scientific reports; it attempted to muzzle James Hansen and other climate scientists; and it rejected all calls to limit America's greenhouse gas emissions. What's more, the administration's intransigence had the effect of blocking international action. After all, if America refused to accept limits, why should China, Russia, Brazil, Indonesia, and other big emitters accept them?

Paul O'Neill, the former CEO of the aluminum company Alcoa who served as Bush's first secretary of the treasury, provided the most damning summary of the Bush-Cheney agenda. Early in his first term, Bush asked O'Neill to draft a plan of action on global warming. But O'Neill's plan was completely ignored. Instead, O'Neill told me later, the Bush administration "cherry-picked" the science on climate change to justify taking no action, "just like it cherry-picked the intelligence on weapons of mass destruction" to justify the invasion of Iraq.

The Tobacco Connection

Not all of the blame for America's foot-dragging can be laid at the feet of Bush, Cheney, and their fellow Republicans, though. Schellnhuber lamented the "lost decade" under Bush, but there were actually two lost decades, and the first occurred while Democrats controlled the White House. When Bill Clinton took office in 1993, he and especially Vice President Al Gore wanted to tackle global warming, but they found the opposition within Washington insurmountable. Congressional Republicans were implacably opposed to any measures that would reduce consumption of fossil fuels, but many Democrats felt the same, not only in Congress but within the Clinton administration itself. Nothing made the point more plainly than the Senate's 95 to 0 vote in 1997 to oppose American participation in any international agreement—that is, the impending Kyoto Protocol—that imposed mandatory emissions reductions on the United States but not on China.

Such powerful bipartisan opposition reflected the fact that curbing greenhouse gas emissions would strike at the heart of America's political economy, not to mention the profits of three of its most powerful industries: oil, coal, and autos. The power of those industries and the reliance of so much of the American way of life on abundant, cheap oil—without it, goodbye, suburbs—helps explain why the United States was slower to address the climate threat than Europe and Japan were, said Everett Ehrlich, who chaired the Clinton administration's interagency deliberations on climate change. "The U.S. is more like an OPEC nation—an energy producer—while the Europeans and Japan are energy consumer nations," explained Ehrlich, Clinton's undersecretary of commerce. "Our natural resource industries are very powerful, and their executives saw dealing with climate change as punitive to their interests. We heard about it repeatedly from them."

The carbon lobby not only complained; it devoted enormous amounts of money and effort to blocking action. For years, America's energy companies had showered politicians with campaign contributions and deployed armies of lobbyists to protect their general interests in Washington. As global warming became an issue in the 1990s, these companies responded by launching a multimillion-dollar public relations campaign aimed at discrediting the science of global warming in the minds of lawmakers, journalists, and the public.

Remarkably, the carbon lobby's attack relied on many of the same tactics and strategies—even the same scientists—that the tobacco industry had previously used to resist government regulation of cigarettes. Just as tobacco companies denied that smoking causes cancer, so the carbon lobby denied that greenhouse gas emissions pose a threat to human well-being. In each case, the companies cloaked their self-serving claims in a mantle of apparent scientific respectability. No man did more to assist them than Frederick W. Seitz, who had begun his scientific career as a young physicist working on the Manhattan Project, America's ultra-secret World War II project to build an atomic bomb.

You could call Seitz the $45 million man. That's how much money he received from the R. J. Reynolds Tobacco Company in the 1970s and 1980s to fund medical research that blunted public understanding of the health effects of smoking. Much of this research took place at Rockefeller University, an institution founded and subsidized by the Standard Oil fortune, where Seitz was president. On top of his Rockefeller salary, Seitz earned $585,000 from R. J. Reynolds for supervising its health research efforts, according to company documents that Seitz confirmed when I interviewed him in 2006. The research Seitz supervised was mainly concerned with medical issues, but it avoided the central health question facing Reynolds. "They didn't want us looking at the health effects of cigarette smoking," Seitz told me. Nevertheless, the research served R. J. Reynolds's purposes, for it enabled the tobacco industry to publish newspaper and magazine advertisements for decades citing its multimillion-dollar research program as proof of its commitment to science—and to argue that the dangers of smoking cigarettes were uncertain.

Or, to quote a tobacco industry planning memo from that time, "Doubt is our product."

"Looking at stress, at genetics, at lifestyle issues let Reynolds claim it was funding real research," explained Stanton Glantz, a professor of medicine at the University of California, San Francisco, and a coeditor of *The Cigarette Papers* (1998), which exposed the inner workings of the Brown & Williamson tobacco company. "But then it could cloud the issue by saying, 'Well, what about this other possible causal factor [for lung disease]?' It's like coming up with fifty-seven other reasons for Hurricane Katrina rather than global warming."

For his part, Seitz told me he was comfortable taking tobacco industry money, "as long as it was green. I'm not quite clear about this moralistic issue. We had absolutely free rein to decide how the money was spent."

I asked whether his research gave the tobacco industry political cover.

"I'll leave that to the philosophers and priests," he replied.

In the 1990s, Seitz began arguing that the science behind global warming was likewise inconclusive and certainly didn't warrant imposing limits on greenhouse gas emissions. He made his case vocally, trashing the integrity of IPCC scientists on the op-ed page of the *Wall Street Journal*; publicly circulating a letter to the Clinton administration, accusing it of misrepresenting the science; authoring papers that said global warming was a fanciful threat devised by environmentalists and unscrupulous scientists pushing a political agenda.

But Seitz was only the highest-ranking scientist among a group of advocates who, beginning in 1991, disputed every suggestion that climate change was a real and present danger. As a former president of the National Academy of Sciences (from 1961 to 1970), he gave such objections instant credibility. But it was the Global Climate Coalition, an organization created and funded by the coal, petroleum, utility, and auto industries, that did the most to promote these views to government, business, and media leaders and thereby to the public at large. Although Ross Gelbspan and other journalists published occasional exposés of the coalition's funding sources and political agenda, the deniers' assertions were generally taken at face value in congressional hearings, news stories, and other public forums and ended up having considerable effect.

"The goal of the disinformation campaign wasn't to win the debate," Gelbspan later explained. "The goal was simply to keep the debate going. When the public hears the media report that some scientists believe warming is real but others don't, its reaction is, 'Come back and tell us when you're really sure.' So no political action is taken."

It was all in keeping with the PR strategy "Doubt is our product."

"They've done a very good job of getting their perspective to receive much more attention than it deserves on the basis of its scientific credibility," James Hansen said of the deniers. The NASA scientist accused them of "acting like lawyers, not scientists, because no matter what new evidence comes in, their conclusion is already decided." As the scientific case for climate change solidified in the 1990s, said Hansen, the deniers' counterarguments shifted accordingly. At first, they denied the earth was warming at all. When that became untenable, they said that any warming would be small and have few ill effects. Next, they said that even if there was warming, human activity wasn't the cause. By the end of the Bush years, they had been reduced to what was their core objection all along: that cutting greenhouse gas emissions would "wreck the economy," as Bush put it.

"Not trivial" is how Seitz reckoned the influence he and his fellow deniers had. Their arguments were frequently cited in Washington policy debates, especially in the lead-up to the 95 to 0 Senate vote prior to Kyoto. The deniers' effect on news media coverage was also profound, said Bill McKibben, who in 1989 published the first major popular book on global warming, *The End of Nature*. Introducing the tenth anniversary edition of his book in 1999, McKibben noted that virtually every week over the past decade studies had appeared in scientific publications painting an ever more alarming picture of the global warming threat. Most news coverage, on the other hand, "seems to be coming from some other planet."

"In the U.S. you have lots of news stories that, in the name of balance, give equal credence to the skeptics," Fiona Harvey, the environment reporter for the *Financial Times*, Britain's leading business newspaper, told me in 2005. "We don't do that here, not because we're not balanced but because we think it's unbalanced to give equal validity to a fringe few with no science behind them."

As of April 2010, much of the U.S. media has still not learned this basic lesson of news judgment. U.S. media coverage of global warming had begun to improve in early 2006, when, in the aftermath of Hurricane

Katrina and emboldened by the release of Gore's documentary *An Inconvenient Truth,* many news organizations finally made it clear that an overwhelming majority of scientists believed man-made global warming is real, already under way, and very dangerous. But the improvement turned out to be short-lived. By late 2009, key parts of the media in the United States and internationally had reverted to their long-standing posture of scientific illiteracy and de facto complicity with the deniers' disinformation campaign.

As the Copenhagen climate summit began in December 2009, almost every major news organization in the world gave front-page coverage to the deniers' unfounded accusations of widespread fraud on the part of leading climate scientists. Quoting people out of context and cherry-picking data, the deniers accused scientists at the Climate Research Unit of the University of East Anglia in Britain of falsifying results and then lying about it, and of conspiring to suppress dissenting views. The only news organization that took the time to investigate rather than merely echo these charges was the Associated Press. A team of AP reporters read and analyzed each of the 1,073 stolen e-mails, a total of about 1 million words of text. The AP found that some of the East Anglia scientists had said nasty things about deniers — hardly a surprise, considering all the nasty things deniers had said about them. Some East Anglia scientists also discussed concealing data, though in the end they did not do so. Bottom line: the AP found zero evidence of fraud, a conclusion later shared by two official investigations by British government bodies. In the words of the AP's headline, "Science Not Faked, but Not Pretty."

But by the time that AP story was published, the rest of the media had embraced the deniers' framing of the controversy as "Climate-gate," thus implicitly endorsing the notion that evil deeds were afoot and amplifying the underlying suggestion that climate science was bunk. A few weeks later, news outlets again advanced the deniers' agenda when they repeatedly devoted ominous headlines to a handful of inaccuracies discovered in the IPCC's *Fourth Assessment Report,* including the mistaken assertion that the Himalayan glaciers could disappear by 2035. That there might be a handful of errors within a three-thousand-page-long report is not surprising, but most stories portrayed it as deeply suspicious. Worse, they failed to explain that these inaccuracies, which the

IPCC acknowledged and corrected, did not undermine the core message of the report or the overall findings of climate science.

One amusing exception to the media's generally poor performance came from Tom Toles, the veteran syndicated cartoonist. In February 2010, he published a cartoon whose first panel showed a man watching television at home while on the screen a talking head announces, "After a comprehensive review of the climate science, we have concluded that climate change is 99.5 percent certain." In the second panel of the cartoon, the talking head adds, "Not 100 percent, as we previously stated." In the last panel, the man at home angrily pumps his fist in the air and shouts, "Aha! I *knew* it." In the lower right corner of the panel, a miniature version of the man adds, "It follows that it's all a hoax."

Crime is a strong word, but it is one used by Schellnhuber, Hansen, Gelbspan, and others angered by the carbon lobby's deceptive campaign to put its financial interests ahead of the future of our children and civilization. As a journalist, it shames me that the lobby could never have succeeded without the assistance of the media; if the deniers themselves committed a crime by misrepresenting the science on climate change, many mainstream news outlets aided and abetted that crime, a journalistic failure as profound as any in modern U.S. history.

Personally, I rarely bother to engage with deniers anymore; it's a waste of time. As a journalist, my credibility depends on being open to new information and changing my views as necessary. Deniers, by contrast, are true believers. They start with their conclusion—global warming is a hoax—and then work backward to assemble the supporting "evidence." Like those who dispute evolution, they are ideologues: their minds are made up and will not be confused by facts that do not fit their agenda. Thus they seized on a regional cold spell that chilled the northeastern United States in December 2009 to mock the very idea of global warming, a stance that only illustrated how little they understood actual climate science. It's pointless to explain to them that global warming does not cancel winters or even rule out individual cold snaps; it only makes winters, on average and over time, shorter and warmer, which is exactly what happened globally in the winter of 2009–10. (Indeed, NASA determined that 2009 was the second-warmest year in the thermometer record and the decade of 2000 to 2009 was the warmest ever recorded.) Nor are deniers swayed by the fact that the United States

National Academy of Sciences, like virtually every other major national scientific academy in the world, has repeatedly declared that man-made global warming and climate change are real and pose profound dangers to society. Hearing that only fortifies their conviction that the climate conspiracy is larger and more nefarious than they realized. The real conspiracy, of course, has been the long-standing disinformation campaign mounted by the giant corporations of the carbon lobby, but I doubt most deniers are aware that they are mouthing talking points originally developed by big money interests. Nor do I expect them ever to change their views. "[Nobel Prize–winning physicist] Max Planck used to say that people don't change their minds [because of evidence]," observed Robert May, the former president of the Royal Society, Britain's national academy of science. "Science simply moves on and those people eventually die off."

The problem is, they may end up taking a lot of us with them. Frederick Seitz and his fellow deniers may look silly on scientific grounds, but they can claim enormous political achievements. For many years, despite the evidence, they managed to make millions of people, including journalists and others who should have known better, question the reality of man-made global warming and climate change. (Just a few days ago, a woman in the wine business—in California, no less—told me that her industry's slow response to global warming was entirely understandable, "since there are some people who believe in it and some who don't.") Most damaging of all, the deniers succeeded in prolonging the Washington policy battle, and therefore global action, long after the issue should have been settled. Thus they delayed actions to reduce greenhouse gas emissions precisely when such reductions would have mattered most. "Had some individual countries, especially the U.S., begun to act in the early to mid-1990s, we might have [avoided dangerous climate change]," Michael Oppenheimer, the Princeton geophysicist, told me. "But we didn't, and now the impacts are here."

The Crime Continues

I recount all this history because it helps explain not only how we got into this mess but also how we might get out of it. The debate on climate change has shifted considerably in recent years, even in the deniers' stronghold of the United States, as more and more people and institutions recognize the urgency of change. But actual change—tangible, far-reaching reforms in how we produce, consume, and organize our economies—lags far behind. That is in no small part because many of the same interests and ideologues remain determined to obstruct progress. The crime, in other words, continues. I believe that the rest of us should respond accordingly: by calling the perpetrators to account, bringing them to justice, and prohibiting them from further imperiling our future and that of our children.

Tobacco companies were eventually made to answer for their crimes in a court of law; the carbon lobby deserves no less. In 1994, Mississippi attorney general Michael Moore filed a lawsuit against the tobacco industry, claiming that its products had caused a health crisis that was costing his state billions of dollars in treatment expenses. Moore's suit sparked similar litigation on the part of forty other states. By 1997, major tobacco companies had agreed to a settlement that required them to pay hundreds of billions of dollars to state governments across the United States to offset the costs of treating tobacco-related illnesses and to finance public education campaigns against smoking. The point of the fine was not only to punish the bad behavior of the past, but also to deter bad behavior in the future.

Some climate activists have urged similar actions against key members of the carbon lobby, especially ExxonMobil—long the most outspoken opponent of climate action and the biggest funder of denier activities. Calling the carbon lobby's disinformation campaign "one of the great crimes of our era," John Passacantando, the former executive director of Greenpeace USA, said he was "quite confident" that class-action lawsuits will be filed against the corporations involved. He told executives from one company, "You're going to wish you were the tobacco companies once this stuff hits and people realize you were the ones who blocked [action]."

Beyond putting the carbon lobby on trial, the larger goal must be to keep the lobby and its intellectual collaborators—the think tanks and spokespeople who spread its message in the public arena—from further distorting society's decision making. These companies and individuals have, through their past actions, forfeited any claim to credibility. They have been wrong—repeatedly, sometimes deliberately, and for the most part unrepentantly—for years. Why should anyone still listen to them? But the media and other public outlets continue to give platforms to deniers, generally without challenging their claims. Media companies also gladly accept millions of dollars' worth of advertising from energy companies such as Chevron, which in 2009 ran ads that blamed individual consumers but not corporate agendas for carbon emissions.

Deniers have a right to express their opinions, but it should not be an unfettered right. The U.S. government prohibits tobacco companies from running cigarette ads on television—why shouldn't it prohibit companies from running misleading ads about climate change? And if deniers wish to testify before Congress, lobby government agencies, appear in the media, or otherwise influence public policy and debate, their audiences should first be reminded of their track record on the issue and the deniers should be forced to defend their unscientific ranting. We don't allow tobacco companies and their apologists to decide public health policy; we shouldn't let fossil fuel companies and their dupes decide climate policy.

True, some companies that initially denied global warming claim to have turned over a new leaf, but few have actually done so. British Petroleum, which was one of the first defectors from the Global Climate Coalition, in 1997, later rebranded itself BP, as in "Beyond Petroleum," to signal its new high-mindedness. The company has boasted of spending $8 billion a year to research and develop low-carbon energy sources, but it spends twenty times that much on traditional oil and gas development. Meanwhile, its obsession with maximizing profits led BP to cut corners on safety, as was horrifyingly demonstrated by the deep-sea gushes that released tens of millions of gallons of oil into the Gulf of Mexico in 2010.

In 2008, even ExxonMobil said it would stop funding denier activities, but it didn't. With the coming to power of President Barack Obama, deniers shifted their critique of climate change from science to economics, with ExxonMobil leading the way. In the opening months of Obama's presidency, Democratic congressmen Henry Waxman of California and Edward Markey of Massachusetts collaborated with the White House to introduce the American Clean Energy and Security Act, which soon became the leading piece of climate legislation on Capitol Hill. The bill aimed to reduce U.S. greenhouse gas emissions by a mere 4 percent from 1990 levels by 2020—well short of the IPCC's call for 25 to 40 percent reductions, much less Schellnhuber's proposed 100 percent cuts. But even Waxman-Markey's goal was too ambitious for ExxonMobil, the American Petroleum Institute, and other fossil fuel interests. Exxon-Mobil funded a study by the Heritage Foundation, a right-wing Washington think tank, which claimed that passing the Waxman-Markey bill would cost millions of jobs, drive energy prices through the roof, and undermine U.S. competitiveness in the international marketplace. Gasoline prices, the study charged, would jump to $4 a gallon. But when API trumpeted this eye-popping claim to the media, it neglected to mention that the Heritage study had been funded by ExxonMobil. Nor did it acknowledge that, even according to the Heritage study, gas prices would not hit $4 a gallon until the year 2035.

ExxonMobil and most other giant fossil fuel corporations are dinosaurs that belong to the twentieth-century energy order. Left to their own devices, they will not abandon fossil fuels anytime soon, certainly not soon enough to avoid catastrophic global warming. So they cannot be left to their own devices.

To the ordinary person, changing the behavior of some of the richest corporations in the world may seem an impossible task, but in fact there are concrete recent examples of organized citizens doing just that—you just don't hear about them on most TV news shows (perhaps because the shows are often financed by the same corporations' advertising). In 2007, a network of citizens' groups across the United States set out to block coal and electric companies from building new coal-fired power plants. Loosely coordinated by the Sierra Club, the Beyond Coal campaign employed a variety of tactics, including legal challenges, public protests, and appeals to elected officials. It worked with a broad array of interest groups, from health professionals worried about air pollution's effects to farmers and ranchers, business and church groups. By June 2010, the campaign had succeeded in getting 129 new coal-fired power plants either canceled or prohibited. Another 51 plants faced legal challenges. Of the 231 plants that had been planned as of 2000, the Sierra Club estimated that only 25 were likely to be permitted for operation. Meanwhile, as Lester Brown of the Earth Policy Institute reported, Wall Street had downgraded coal company stocks, and prominent national politicians, including Senate majority leader Harry Reid and the governors of California, Florida, Michigan, and Washington, had expressed their opposition to building more coal plants.

Impressive as such direct action can be, however, it will not suffice. Stopping climate-destructive behavior is only half the battle. If we are to decarbonize our societies rapidly enough to avoid catastrophic climate change, we must also fight *for* things. We must push for rapid deployment of low-carbon alternatives, not just in the energy sector but in agriculture, construction, and across the entire economy, in rich and poor countries alike. This will require fundamental changes in government policies, many of which now perversely encourage climate-destructive behavior.

The rules that govern energy, agriculture, and other economic sectors will not be changed simply because reform is necessary to preserve a livable planet. Genuine reform in all spheres of public policy is usually the result of governments being pressured from below by determined, mobilized citizens. The media and even some environmental groups often give the impression that the best way for people to fight climate change is through individual lifestyle changes—recycle more, drive less, eat less meat. Lifestyle adjustments are important, but the real key to shifting our civilization's climate trajectory is to change the governmental policies that shape the decisions that all of us, consumers and corporations alike, make. That means that politics must be committed.

Defeat of the Dragons

I talked earlier in this book about the inspirational power of fairy tales, but I've come to realize that the fairy tale I most want to read to Chiara hasn't been written yet. I would call it *Defeat of the Dragons.* The story would be set on a beautiful island where the weather has started to get weird. Temperatures are much hotter. Rains aren't falling when they should. Water is growing scarce, making it difficult to grow enough food. Islanders are getting nervous, some are suffering, a few are suffering quite grievously. The learned of the island—call them the wizards—begin to study the problem and come to the conclusion that what is changing the weather are the dragons.

There are not many dragons on the island, but they are large, powerful creatures that are used to getting their own way. Everyone on the island, even little children, knows that the dragons don't smell good. They have bad breath, and their farts stink something terrible. In the past, the islanders saw no other choice but to put up with the bad smells: after all, the dragons were big and sometimes mean. Now, however, the wizards have determined that the dragons' foul odors are not just gross but dangerous: they are trapping the sun's rays on the island, which is why the weather is getting weird.

The wizards want the dragons to change their diets: if they eat less meat and more fruits and vegetables, their emissions won't be so noxious. The dragons refuse. They like things the way they are, they actually think their breath and farts smell good, and they don't care what others think. The islanders try everything they can think of to change the dragons' minds. They point to the weird weather. They cite studies. They invoke the dragons' self-interest, pointing out that an overheated island will not be good for the dragons' own offspring. Eventually, they appeal to the island's governors, the In-Chargers, asking them to *make* the dragons change. Nothing works. The dragons respond by deriding the wizards' knowledge, sweet-talking and bribing the In-Chargers, and eating more meat than ever.

Finally, the islanders come to believe it's no use trying to convince or bargain with the dragons. The only hope is to defeat them: to force them to stop emitting the gases endangering their island or to die trying.

At this point in a classic fairy tale, a hero would emerge to lead the battle against the dragons. In *Defeat of the Dragons,* however, there would be not just one hero but thousands. Each would work in his or her own way while collaborating with comrades to advance the cause. Many of the heroes would be children, and they would devise the masterstroke that saves the day. Maybe they would borrow a trick from *The Emperor's New Clothes* and deflate their foes with embarrassing wit. The children could shout what everyone else knows but has been afraid to say: "Dragons, we have shared this island with you for a very long time, but the time has come to tell the truth: you have really bad breath and very stinky farts. Your behavior is making the rest of us sick, and if you don't change your diets, we will have to take your food away."

A battle ensues, and after some fierce clashes the dragons, amazingly, go down to defeat. The dragons that are captured are given a chance to adopt a new diet, heavy on vegetables and fruits, which to their surprise they find they prefer. The weather, alas, stays weird for a long time. But the islanders get better at coping with it, and by the time their children are old enough to become grandparents, the climate has found a fresh equilibrium, a new normal. Though quite different from its original state, the island is still beautiful in its way; it is, after all, home.

Well, *Defeat of the Dragons* needs some work, but you get the idea. That's the fairy tale I'd like to read to Chiara someday soon. In fact, it's the fairy tale I'd like to see her help bring about when she gets a little older. Until then, it's up to the rest of us.

Yes We Can

Leading the fight against global warming amounts to a heroic quest, and when we spoke in Santa Fe, Schellnhuber told me Barack Obama was just the man for the job. "This is the hour of leadership," the German scientist said. "Nobody else has the stature to go to Copenhagen and convince governments to launch a Green Apollo program. Obama has an opportunity to become the Abraham Lincoln of our time. We remember Lincoln for abolishing slavery. We would remember Obama for saving the world from climate catastrophe."

This looked entirely possible the night Obama was elected president. I was lucky enough to be in his hometown of Chicago on Election Day 2008 and to witness his victory speech in Grant Park. As the overflowing crowd waited for him to arrive, the faces around me—a diverse mix of black and white, young and not-so-young—wore happy but dazed expressions, as if people sensed that something amazing was happening but couldn't yet believe it was real. Then Obama took the stage, flashing that incandescent grin and holding hands with his wife and two young daughters, and the crowd's dawning sense of euphoria exploded into a roar heard 'round the world. People everywhere knew this was a historic moment, but in Grant Park you felt it viscerally, as if an electric current were passing from one person's body to the next: the United States, a nation built on slavery but promising justice for all, had just elected its first nonwhite president. If that was possible, it seemed anything could happen.

In his speech Obama made the point explicit. Telling the life story of 106-year-old Ann Nixon Cooper, who had voted for him that day in Atlanta, Obama maintained that "unyielding hope" had triumphed over long odds again and again in the past. When Ms. Cooper was born in 1902, Obama said, someone like her could not vote because of both her sex and her skin color. But she lived to see many supposedly impossible things come to pass: women and African Americans winning the right to vote, a man walking on the moon, the fall of the Berlin Wall. "Yes we can," Obama had said during his campaign, urging Americans to believe in their ability to come together and change their country. Now unyielding hope had made a man of mixed race and humble origins the nation's forty-fourth president, and he hoped it would continue to guide America through the undoubted difficulties ahead. For "while we breathe, we hope," Obama concluded. "And where we are met with cynicism and doubt and those who tell us we can't, we will respond with that timeless creed that sums up the spirit of a people: Yes we can."

If ever a problem cried out for the unyielding hope of "Yes we can," it is climate change, and President-Elect Obama seemed to agree. In his victory speech he called "a planet in peril" one of the three biggest problems awaiting him in the White House, giving it equal billing with "the worst financial crisis in our lifetimes" and the wars in Iraq and Afghanistan—the first time an American president had put climate change so high on his agenda. What's more, Obama seemed inclined toward the very sort of Green Apollo project that Schellnhuber had advocated. During the campaign, candidate Obama had proposed that the government make large investments in alternative energy in order to protect not only the atmosphere but America's economy and national security by reducing its dependence on foreign oil. As president, Obama carried through on this commitment in his first weeks in office, embedding green energy initiatives throughout the economic stimulus package intended to halt the free fall of the economy after the global financial crisis. The stimulus package contained $71 billion in direct green spending and $20 billion in green tax incentives. Spending on alternative energy increased to three times the level it had been under Bush.

Greening the stimulus package was a good start, but it paled against all the United States needed to do if the climate is to stabilize in time. Indeed, the urgency of the challenge facing Obama stemmed largely from how late he was coming onstage in the climate drama. The two lost decades preceding his arrival in the White House meant that his policies, though a substantial improvement on Bush's, fell well short of what was needed to limit global temperature rise to 2°C. By 2009, preventing catastrophic climate change would require almost a revolution in America's behavior. It would require, as Schellnhuber said, a Green Apollo program.

In urging such a program, Schellnhuber had unwittingly put a new name on a set of ideas and policies that Al Gore and numerous others had already proposed. In his 1992 book *Earth in the Balance,* Gore had urged the U.S. government to initiate a green Marshall Plan; like the aid program that Washington financed after World War II to resuscitate the war-torn economies of Europe, Gore wrote, a Green Marshall Plan would help developing nations shift from fossil fuels to renewable energy sources. In my 1998 book *Earth Odyssey,* I proposed a Global Green Deal, modeled on the New Deal U.S. president Franklin Roosevelt had championed against the Depression of the 1930s. Six years of traveling around the world had convinced me that the fight for environmental solutions had to be linked to the fight against poverty: you cannot ask poor people to do without today in order to build a greener tomorrow, and when half the world is poor, the poor cannot be ignored. In 2008, *New York Times* columnist Thomas Friedman argued in his book *Hot, Flat, and Crowded* that America needed to launch a Green Revolution, both to combat climate change and to free the country from the grip of what he called "petro-dictators."

By 2009, versions of a Global Green Deal had been urged by senior politicians and key citizens' groups the world over, including the secretary-general of the United Nations, the prime minister of Japan, the foreign minister of Germany, and leading labor and environmental groups in Britain and the United States, including the latter's Apollo Alliance. In February 2009, at the depth of the financial crisis, Gore and UN secretary-general Ban Ki-Moon urged governments to stabilize the situation with spending that not only addressed social needs but launched "a new green global economy." Noting that thirty-four nations planned $2.25 trillion in stimulus spending, Ban and Gore warned that channeling the money "into carbon-based infrastructure and fossil-fuel subsidies would be like investing in sub-prime real estate all over again."

I've come to believe that Green Apollo is the best term for what we need: a crash program, in rich and poor countries alike, to jump-start the transition to an economy that is both climate-friendly and climate-resilient. By harkening back to the United States space program, Green Apollo invokes a historic achievement of which all humanity can be proud, thus calling forth the elevated sense of purpose that will be essential to confronting climate change. Green Apollo also conveys the short time frame—ten years, like the race to the moon—we face. Green Apollo will have to mobilize public and private resources in all countries with a wartime sense of urgency. Relying on a mixture of government policies and market mechanisms, Green Apollo programs must encourage rapid deployment of green technologies and practices, especially in the fields of energy, transportation, construction, forestry, and agriculture. Above all, Green Apollo must establish a rising long-term price for carbon, thereby channeling the enormous power of the market—of the purchasing power of businesses, governments, and consumers the world over—toward actions that reduce rather than increase our collective carbon footprint. Green Apollo programs will also shift government subsidies away from practices that make climate change worse, such as the United States government's copious financing of highways and oil and coal projects, and toward their green counterparts. Green Apollo will also encourage adaptation—the installation of sea defenses, efficient water systems, and other measures to reduce communities' vulnerability to the climate impacts that can no longer be avoided.

How to pay for all this? The good news is that Green Apollo programs, done properly, promise to green our wallets as well as our societies. To be sure, these programs will cost money to set in motion, but they will pay for themselves over time. Sustained investment in energy efficiency, tree growing, mass transportation, wind turbines, solar power, soil rejuvenation, smart electrical grids, and kindred green technologies will generate a torrent of new jobs, profits, tax receipts, and innovation while reducing energy waste and avoiding the expense (and human suffering) that would result if we failed to install protections against floods, drought, and other climate impacts. Last but not least, Green Apollo programs will help lift millions of people around the world out of economic distress—a vital priority in its own right but also a prerequisite to gaining poor and emerging countries' collaboration in reducing greenhouse gas emissions before it's too late.

What would a Green Apollo project look like in practical terms? Describing it in detail would require a whole separate book, and its specific features will vary depending on the country pursuing it. The following few pages sketch only its general principles. I will focus on the United States, partly because it's the place I know best, but also because the United States—the leading greenhouse gas emitter and the world's richest, most technologically advanced nation—has the greatest influence over whether our civilization will do what is necessary to avoid the unmanageable and manage the unavoidable of climate change.

The Green Apollo Program, Explained

Luckily, although the hour is very late, we already know much of what needs to be done. Success stories abound, both about how to slash greenhouse gas emissions and how to adapt to climate impacts. For their adaptation policies, Green Apollo programs should draw on the lessons of innovators featured in this book—Ron Sims of King County, Washington; Chris West of UKCIP; Pier Vellinga and Aalt Leusink of the Netherlands; Ivor van Heerden and Hassan Mashriqui of New Orleans; Saleemul Huq of Bangladesh; Yacouba Sawadogo of Burkina Faso—and many others. Green Apollo programs should both publicize these stories so others can emulate them and help bring them to scale—for example, by disseminating information through government outreach mechanisms (presidential speeches, constituent meetings, agricultural extension agents, and so on) and by collaborating with the news media and nongovernmental organizations. We do not need to reinvent the wheel; we know what many of the best practices are. What's needed is to apply those practices as quickly and widely as possible.

On the mitigation front, some of the most important lessons come from Amory Lovins, the president and chief scientist of the Rocky Mountain Institute in Colorado. For decades now, Lovins has been demonstrating how businesses, governments, and society as a whole can shrink environmental footprints while maintaining and even increasing economic well-being. The assumption that burning less oil and coal will mean higher prices, fewer jobs, and lower living standards has long been the heart of the argument against cutting greenhouse gas emissions, endlessly repeated in Washington, Copenhagen, and beyond. But it is a myth, says Lovins, an Oxford-trained physicist with a walrus mustache who has collaborated with scores of corporations, governments, and institutions around the world, including the U.S. military, over the years. Lovins argues not from economic theory but real-world experience. Writing shortly after the Copenhagen climate summit, he noted that "many business leaders understand . . . that energy efficiency is one of the highest-return and lowest-risk investments in the whole economy. Dow Chemical has saved $9 billion by investing $1 billion in [better] energy efficiency. DuPont made billions by cutting its greenhouse-gas emissions 72 percent during 1990–2004, and is now expanding that cut by another 15 percent." Look again at the numbers Lovins cites: DuPont's 72 percent emissions reductions are roughly double the 25 to 40 percent cuts the IPCC has urged. They even approach the 100 percent cuts advocated by Schellnhuber—evidence that we can meet the mitigation challenge, if we choose to do so.

Because energy efficiency is so lucrative, it should be a cornerstone of Green Apollo programs the world over, even in developing countries. That may sound counterintuitive. Being poor, such countries need to use *more* energy in order to rise from poverty. But improving efficiency is still the best way to acquire it. The reason is that energy systems in developing nations are even more plagued by waste than their counterparts in wealthy nations are; developing nations' poverty has led them to rely largely on old, inefficient motors, furnaces, and related technologies. China is a good example. Much of its energy infrastructure was inherited from its former ally the Soviet Union; when I visited in the late 1990s, experts joked that China's notoriously polluting power plants had been built in the 1950s with Soviet designs from the 1920s. Ironically, that inefficiency gave China huge opportunities to reduce pollution without sacrificing economic output. Studies supervised in the 1990s by Zhou Dadi, a top government climate adviser, showed that China could use 40 to 50 percent less energy if it installed currently available energy technologies, such as more efficient refrigerators, light bulbs, and air conditioners; insulation for China's perpetually drafty buildings; smarter electric motors and equipment. Recent research by Jiang Kejun, a top government climate adviser at the Energy Research Institute, has shown even greater potential: a strong investment in efficiency would enable China's carbon emissions to peak by 2030.

Green investments are also excellent job creators. They tend to be more labor-intensive than capital-intensive—that is, they rely more on workers than on equipment. Germany's program of upgrading the energy efficiency of its oldest housing stock is a prime example. Replacing old furnaces and installing efficient windows and lights produced thousands of well-paying laborers' jobs that by their nature could not be outsourced; many of these jobs were created in the former East Germany, where unemployment was approaching 20 percent when the program began in 2005. By 2008, Germany had retrofitted or constructed anew more than 800,000 apartments. "The program has proven to be a genuine motor for economic growth, employment and innovation in the construction industry—a win-win situation for business, society and the climate," said Wolfgang Tiefensee, then Germany's minister of transport, building, and urban affairs.

Energy efficiency can take us a very long way toward a climate-friendly future, but new technologies will also be needed. Saving energy alone will not suffice on a planet whose human population is growing in both its raw numbers and its appetites. If air travel, one of the most carbon-intensive activities on earth, is to continue at anywhere near current levels, a revolutionary breakthrough in how planes are fueled is needed. If humans are to increase our reliance on organic agriculture, as seems essential on both climate mitigation and adaptation grounds, we need to answer the question posed by Ma Shiming of the Chinese Academy of Science: can organic match the production yields of industrial agriculture, and if so, how soon?

Improved efficiency—in how we use not only energy but water and other natural resources—can play a critical role here as well. Because saving energy is both the fastest and the most lucrative way to slash greenhouse gas emissions, investing in efficiency can buy our societies time to develop the breakthrough technologies needed to complete the transition to a climate-friendly economy. Efficiency savings can also furnish part of the financing needed to bring these technologies to scale throughout our societies.

Nuclear power is often cited as one of the technologies that must be expanded to combat climate change, but the superior speed and economics of energy efficiency leave nuclear hopelessly behind. As it happens, I first heard the term *global warming* from a nuclear industry executive while I was researching my first book, *Nuclear Inc.* Speaking in the early 1980s, the executive assured me that, despite a stall in reactor orders in the wake of the Three Mile Island accident, his industry had a bright long-term future. As humanity approached the turn of the century, he explained, the drawbacks of coal-fired power plants, including their contribution to global warming, would reacquaint governments and citizens alike with the advantages of nuclear. Of course, many outsiders continue to believe that nuclear power poses unacceptable safety and security risks. After all, there is still no safe disposal method for nuclear waste, and nuclear weapons proliferation is a danger of the highest order; every nuclear weapons state in the world began its program by exploiting the civilian applications of fission. But put aside those concerns for the moment. As my research in the industry's own libraries revealed, nuclear power has been plagued since the industry's birth in the 1950s by the tremendous economic expense of developing and building nuclear power plants. Indeed, the technology was commercialized only thanks to gargantuan government subsidies that continue to this day.

Government subsidies are not inherently evil—we'll need plenty of them as we build our new, low-carbon societies—but they must be spent wisely. What makes energy efficiency vastly superior to nuclear power as a tool against climate change is that it saves energy much more quickly and cheaply than nuclear can produce it. Real-world experience shows that nuclear power, as Lovins writes, "saves between two and twenty times less carbon per dollar, twenty to forty times slower, than investing in efficiency and micro-power." In a world of scarce capital, he adds, investing in nuclear power actually makes climate change worse by diverting resources from better solutions.

Lovins enumerated many of those solutions in a book he wrote with the Pentagon called *Winning the Oil Endgame,* which showed how to eliminate oil consumption by the United States by the 2040s at a cost of $15 per barrel (in 2000 prices). Coal, he argues, can also be phased out. Here again, boosting energy efficiency is key, but so is the rise of what Lovins calls "distributed renewables" and "micro-power" technologies. Instead of building large, centralized power plants whose electricity must be transported many miles to its end users, distributed renewables and micro-power technologies are located on-site or nearby. A prime example is cogeneration. Rather than expelling the heat produced to run a building's furnaces and machines, cogeneration recycles it into reusable energy. Wind power and small-scale solar can also displace coal—indeed, they already are doing so. In 2007, despite six years of fossil fuel boosterism by the Bush administration, the United States added more wind power in twelve months than it had added coal power during the previous five years. In 2008, the world as a whole invested more in renewable power plants than in fossil fuel electricity generation. "This revolution already happened," wrote Lovins, adding cheekily, "sorry if you missed it!" And it happened for market-based reasons, he added: ". . . these decentralized competitors make cheaper electricity, [are built] faster, and have less financial risk than big, slow, lumpy power plants, so they can better attract private capital despite their generally smaller subsidies."

Misguided subsidies are one of the most important government obstacles that Green Apollo programs must overcome. Approximately two-thirds of U.S. government energy subsidies have historically gone to fossil fuels and nuclear power, while efficiency, solar, wind, and other renewable sources have received relative pittances. Similar disparities have long distorted transportation policy, with highways grabbing the lion's share of subsidies while mass transit shrivels. Globally, governments provide an estimated $300 billion worth of energy subsidies every year, the vast majority of them to carbon-based fuels. The same bias infects most rich nations' foreign aid programs. Green Apollo would reverse these patterns, dramatically increasing support for renewables, efficiency, and mass transit. (It's worth noting that the leaders of the Group of Twenty nations—the twenty largest economies in the world—agreed in October 2009 to Obama's suggestion to phase out all government subsidies of fossil fuels, an extraordinary shift if it actually comes to pass.) Government subsidies also need to be overhauled in agriculture, forestry, transportation, housing, and other fields to encourage the fastest possible shift to greener practices.

But the single most powerful green tool at government's disposal is its own purchasing power. In the United States, for example, "the federal government is the world's largest consumer of energy and vehicles and the nation's largest greenhouse gas emitter," observed Christian Parenti in a special issue of *The Nation*, which I guest-edited following BP's Deepwater Horizon oil disaster. If state and local government is included, Parenti added, government spending accounts for a whopping 38 percent of the U.S. gross domestic product. Shift all that purchasing power towards clean energy, electric vehicles, and efficient buildings and it would "drive down marketplace prices sufficiently that the momentum towards green tech would become self-reinforcing and spread to the private sector."

Meanwhile, goverments in all nations must also establish a price on carbon. At the moment, carbon emitters are allowed to pollute the atmosphere free of charge. Putting a price on such pollution would give all economic actors an incentive to shift to more fuel-efficient vehicles or invest in solar or wind power. Investors, in turn, would respond to these behavioral shifts by bringing to market goods and services that reduce emissions even further. "We believe we're about to unleash the greatest technological revolution the world has ever seen," Representative Ed Markey declared at the Copenhagen climate summit as he praised the climate bill he and Representative Henry Waxman cosponsored to raise the carbon price. "[This revolution] will be like the telecommunications revolution of the 1990s, when we went from initially having personal computers in a relatively few households to where Google later became part of our very language."

Governments can raise carbon prices in any number of ways, but the two most commonly discussed are taxes and caps. Either will raise the price of carbon. A tax does it directly (just as gasoline taxes already do). A cap does it indirectly; limiting the amount of carbon an economy can emit in effect reduces the supply, which drives up the price. Under a cap system, the government issues permits to emit carbon. To drive continuing emissions reductions, the government issues fewer and fewer permits over time.

Cap-and-trade is the best-known, and most controversial, cap system. Under cap-and-trade, emissions permits can be traded on the open market; this is intended to give companies an incentive to reduce emissions as fast as possible, since they can sell any permits they don't use to companies that need them to meet the cap. Cap-and-trade has succeeded in the past but also failed. It succeeded in the United States in the early 1990s, when it delivered a dramatic reduction in emissions of sulfur dioxide, the pollutant that causes acid rain. It later failed to reduce carbon dioxide emissions in Europe, however, largely because most of the permits were given away rather than sold. That giveaway ended up removing most of the incentive for polluters to reduce their emissions.

Some lawmakers defend the giveaways by arguing that high energy prices will hurt poor, working, and middle-class consumers, but there is a ready solution to that problem. It's called cap-and-dividend. The key difference between cap-and-trade and cap-and-dividend is who benefits from selling the permits. Under cap-and-trade, the government keeps the revenue generated by selling permits; it can use it to fund green energy research but could just as well devote it to pork-barrel projects (as Waxman-Markey seemed likely to do). By contrast, under cap-and-dividend the revenue goes back to the public every month in the form of dividends, sent directly to all citizens equally. These dividends would help people cope with higher energy prices as well as shift to low-carbon alternatives. As Peter Barnes, the California economist and businessman who developed the idea, explained in *Scientific American:* cap-and-dividend "is simple to understand and administer . . . [and] it creates a virtuous circle, in which how people fare depends on what they do. The more carbon any company or individual burns (directly or indirectly), the more that company or individual pays. Because everybody gets the same amount back, people gain if they conserve and lose if they guzzle. This is fair to all, and the poor come out ahead because they burn less carbon than other people do."

Clearly, Green Apollo programs will require bold government leadership, but they will depend just as much on the resources and expertise of the private sector and civil society. Like Kennedy's race to the moon, a Green Apollo program is a mission for the entire nation, coordinated by government but drawing on the creativity, resources, and moral support of society as a whole. Just as the most practical and creative things done in New Orleans to recover after Hurricane Katrina came from civil society, so will Green Apollo programs flourish only if community organizations, churches, schools, and other grassroots institutions get involved, make things happen, and push government to do the right thing. Green Apollo programs should encourage individuals, businesses, and governments at all levels to evaluate current practices according to the two imperatives of the second era of global warming summarized by UKCIP's Chris West: "Are you choosing the lowest-carbon energy source available? And are you building infrastructure that will be resilient to twenty-first-century climatic conditions?"

An Efficiency, but Also a Sufficiency, Revolution

Having outlined the basic principles of Green Apollo programs, let me freely acknowledge the main difficulty they face: many of the initiatives that should animate Green Apollo programs run afoul of the status quo. Our current economic practices may be bad for preserving a livable planet, but they deliver huge profits to ExxonMobil, BP, Massey Energy, and other powerful corporations. They provide jobs and economic activity to coal regions from West Virginia to Shaanxi, China. They enable consumers the world over to inhabit suburbs, enjoy imported strawberries in winter, and jet off to tropical vacations. All these practices will have to change, which in turn will spark resistance from those they benefit. But there is no sense in pretending that the rapid transition we need to make to a climate-friendly and climate-resilient civilization will not involve cost and sacrifice. It will take political courage, diplomacy, and wisdom to navigate this transition — to smooth the transition to alternative livelihoods for coal miners, to compel giant corporations to leave fossil fuels in the ground, to induce individuals to make greener personal choices.

Beyond that, let me be clear that energy efficiency is not a cure-all. By itself, it cannot solve our energy and climate problems. Indeed, if we do not avoid a common trap, it could make them worse. The trap, well explained by George Monbiot in his book *Heat*, is that higher efficiency can actually lead, over time, to *increased* consumption. As economists Daniel Khazzoom and Len Brookes have postulated, when increased efficiency lowers the cost of a given activity, the lower cost provides an incentive to do more of the activity, which can then increase total consumption. For example, if better insulation lowers the cost of heating your house, you might turn the furnace on when you feel a chill rather than put on a sweater. This theory, Monbiot argues, helps explain why the world's energy consumption has risen steadily over the past 150 years even as its energy efficiency has improved by about 1 percent a year.

Economists like to say there is no such thing as a free lunch, and improved efficiency is a good example. (And not just in relation to energy:

water experts tell similar stories about irrigation—when better efficiency lowers the cost of irrigation, farmers expand the amount of land irrigated and end up using more water than previously.) If improved efficiency is to help us create low-carbon economies—and it must; we can't succeed without it—it must not become an excuse for increasing production and consumption. Otherwise emissions simply will not fall as rapidly as required. If efficiency were to increase by 50 percent over the next twenty to thirty years but GDP rose by 2.5 percent a year, within twenty-five years "we'd be back where we are now," according to *Growth Isn't Possible,* a report by the New Economics Foundation in London.

"We need an efficiency revolution but also a sufficiency revolution," argues Wolfgang Sachs, a cofounder of the Wuppertal Institute in Germany and one of Europe's leading sustainability thinkers. "High-efficiency cars that travel one hundred miles on a gallon of gas are useful, but must they also be able to travel one hundred miles per hour?" Even more important, Sachs continues, is to make cars unnecessary in the first place. Echoing the antisprawl arguments of Ron Sims in King County, Sachs urges creating walkable communities and investing in mass transportation so people don't require cars to go about their daily lives.

All this brings us to two great unmentionables in most discussions of coping with climate change: the need to curb both human appetites and human numbers. On a finite planet, we simply cannot expect to increase our levels of material consumption and our populations indefinitely. Of the two, reducing consumption levels is by far the more important, though rarely is this understood in affluent countries. Whenever I give public talks or appear in the media in Europe or the United States, I am invariably asked about the need to reduce population, by which the questioner generally means population in developing countries. And population growth certainly does have environmental impacts, especially within those countries themselves. But globally, the impacts of the high-consumption lifestyles common in rich countries are far greater. For example, an average child born today in the United States will emit 20 gigatons of carbon a year, roughly twenty times more than a child born in Burkina Faso will. To avoid misunderstanding, let me say clearly that it is still important to lower population growth rates in poor countries—but it is important mainly for the sake of those countries themselves, for it reduces the pressure on water, land, and other natural resources; lessens the difficulty of providing schools, jobs, health care, and other services; and probably raises the status of girls and women, since that is by far the surest means of reducing birth rates. But these facts should not obscure a larger truth: globally, it is far more important to reduce the environmental footprint of the affluent, including those who happen to live in poor countries.

Reducing consumption need not involve deprivation. Remember the lament of Rohit Aggarwala, the official leading New York City's efforts to reduce greenhouse gas emissions? The mitigation challenge got steeper year by year, Aggarwala said, as more and more New Yorkers bought energy-sucking flat-screen televisions. Government can and should require manufacturers to maximize the energy efficiency of TVs and other appliances, but consumers have to do their part too by not necessarily purchasing every new gadget corporations try to sell them.

Consumerism—the constant thirst to consume ever more material goods and services—is the secular religion of our time. Propelled by relentless advertising, it is deeply ingrained, especially in rich countries. But it imposes huge environmental costs, both in the harvesting of natural resources used to produce those goods and services and in the pollution they go on to generate. The environmental writer Bill McKibben has gone so far as to call consumerism the main obstacle to tackling climate change, and the problem is by no means confined to the United States. "In the UK, the efficiencies of our transportation and household appliances have increased a lot in recent years, but the country's greenhouse gas emissions have actually risen because our per-capita consumption has grown even faster," said Simon Retallack, a senior fellow at the New Economics Foundation. "To reverse climate change we need to confront the issue of growth, but we're in collective denial about it."

So Much for Hopenhagen

As I rushed through the airport after the collapse of the Copenhagen climate summit, my eye was caught by a large wall photo of Barack Obama. Something about it wasn't right. I was bleary-eyed after a late night of covering the dueling press conferences of the summit's final hours, and it took me a moment to see what was off. Only when I read the accompanying text did I notice that this Obama had a head of lightly gray hair. "Barack Obama 2020," the text said, followed by a quote: "I'm Sorry. We Could Have Stopped Catastrophic Climate Change . . . We Didn't." Sponsored by Greenpeace, the Obama ad—and similar ones featuring the faces of other world leaders—had been erected prior to the summit as an exhortation to reach an ambitious, fair, and binding agreement. Now, the ads read less like an exhortation than a prophecy.

A couple of weeks after the summit concluded, the top negotiator for oil-rich Saudi Arabia told the BBC that his country was "satisfied" with the summit's outcome. That's about all you need to know to judge how close Copenhagen took us toward a climate-friendly future. The summit did not produce the fair, binding, ambitious treaty so many yearned for. Instead, it yielded a high-profile side deal, put together at the last minute by a handful of the world's biggest greenhouse-gas-emitting nations, including the United States and China, the two climate superpowers. This side deal was then very grudgingly endorsed by the European Union and other rich industrial nations and accepted even more reluctantly on the last day of the conference by many, but by no means all, developing nations. The full summit explicitly declined to approve the so-called Copenhagen Accord in its final deliberations. Rather, it voted merely to "take note" of it.

No surprise, really: the side deal was in substance all but toothless, and the United States and other powers imposed it in a take-it-or-leave-it fashion. Contrary to early news reports, the side deal did not pledge to limit global temperature rise to 2°C over the preindustrial level; it merely "recognize[d] the scientific view" that the increase should be kept to 2°C. Worse, the Copenhagen Accord did little to bring this result about. It neither enumerated nor prescribed binding limits on the emissions that drive global warming; it only committed both developed and developing nations to "take action" to "achiev[e] the peaking of global and national emissions as soon as possible. . . ." Emissions reductions would remain purely voluntary, and failing to achieve them would result in no penalties.

There was plenty of blame to go around. The Obama administration's refusal to offer more than 4 percent emissions cuts by 2020 was seen by many other countries, rich and poor alike, as evidence that the

United States under Obama was not that different from what it was under Bush. In his speech to the summit, Obama phrased it differently, of course. He said that the United States would cut its emissions by 17 percent by 2020. But Obama was moving the goalposts. By employing a baseline of 2005, rather than the international standard of 1990, the president made his proposed emissions cuts look much larger than they actually were. It was like promising to kick a forty-yard field goal from the ten-yard line. Perhaps he thought he could get away with it because this kind of subterfuge had become standard practice back in Washington. The authors of the Waxman-Markey climate bill had also congratulated themselves on the bill's stated goal of cutting emissions by 17 percent by 2020 without seeming to recognize that shifting the baseline to 2005 meant they were really aiming to reduce emissions by only 4 percent compared to 1990 levels. In fact, at a press conference in Copenhagen, Representative Waxman went so far as to assert that his bill was "completely consistent" with limiting global temperature rise to 2°C, which wasn't even close to true. I don't think Mr. Waxman was lying, only stunningly ill informed. Like President Obama and so many others in Washington, Waxman was judging the boldness of a given proposal according to its prospects on Capitol Hill more than by its relationship with scientific reality. The problem is, the laws of physics and chemistry, unlike those of humans, do not compromise.

The other climate superpower was not much better. China dragged its feet throughout the Copenhagen summit, resisting calls to accept even long-term limits on its emissions and pressuring poor nations to toe its diplomatic line or risk the loss of development aid. Immediately afterward, China came under harsh criticism for allegedly having sabotaged the summit. Ed Milliband, the climate secretary of Great Britain, charged that China's veto had prevented the Copenhagen Accord from agreeing that rich industrial nations would cut emissions by 80 percent from 1990 levels by 2050 while the world as a whole, including rich industrial nations, would cut emissions by 50 percent by 2050.

But there was method to China's madness, for those proposed emissions cuts violated the per-capita principle. If the world was cutting emissions by 50 percent and rich nations were cutting theirs by 80 percent, the unstated effect was to limit the future emissions of poor and emerging nations. Not only that, but the limits on poor and emerging nations, when calculated on a per-capita basis, fell well short of the per-capita emissions rich nations were allowed. This, I was told by one expert on Chinese climate policies, Beijing would never accept.

Mark Lynas, who was in Copenhagen serving as an unpaid science adviser to the government of the Maldives, rejected such rationalizations as a recipe for global catastrophe. "The historical responsibility argument makes sense in one way only: as an argument for adaptation financing," he told me. Historical responsibility "is not an argument for others to pollute just as much. . . . That is the logic of 'mutually assured destruction'—where human concepts of equity triumph over the necessity for planetary survival."

But the historical responsibility argument cannot be so easily dismissed. As Schellnhuber's WBGU study pointed out, humanity has a limited amount of carbon it can emit over the next fifty years if it wishes to stay within the 2°C limit. The main reason the amount is so limited is that rich industrial nations have emitted so much carbon over the last two hundred years. In effect, they have already occupied most of the planet's atmospheric space, and they came to enjoy prosperous, comfortable lives in the process. To say now to poor and emerging nations that they must limit their emissions because there's not enough atmospheric space left cannot help but offend and anger them.

Hence the logic of the Germans' per-capita budget approach: people in rich and poor countries alike should have a right to the same amount of atmospheric space. That, in turn, means that rich countries must reduce future emissions much more sharply than poor and emerging countries. (Indeed, as Schellnhuber pointed out, some poor nations, such as Burkina Faso, can even increase their emissions slightly.) Furthermore, the largest emitters will have to pay the smallest to finance the latter's shift to alternative energy sources. Without such financial help, coal and other fossil fuels simply remain too cheap for poor and emerging economies to shun.

Rich nations have rejected the idea of paying the poor to limit emissions since the UN Earth Summit in 1992. Now they have no real choice: rich nations themselves cannot survive climate change unless China, India, and other emerging economies constrain their emissions. Instead of resisting, the rich should make a virtue of necessity: helping poor and emerging economies to go green at maximum speed will open export markets for what is shaping up as one of the biggest industries of the twenty-first century.

Launching Green Apollo programs in rich and poor countries alike will be expensive, there's no denying it, but it is silly to say our civilization cannot afford it. Governments found trillions of dollars virtually overnight to pour into the world's banking systems in 2008 and 2009 to stave off financial collapse. As Nicholas Stern has observed, the unraveling of the earth's climate system would carry even graver consequences; surely it deserves a commensurate amount of bailout money. Much of the necessary investment could be found by redistributing funds from bloated military budgets, especially in the United States. The Iraq war alone will cost U.S. taxpayers an estimated $3 trillion, according to Nobel Prize–winning economist Joseph Stiglitz. And Washington spends approximately $250 billion a year maintaining U.S. military bases overseas, in part to make sure that foreign oil continues to flow securely back to the United States. Noting that the taxpayers in his home state of Vermont paid $150.6 million in 2009 for "oil-related military efforts," Bill McKibben pointed out that "if we'd spent that money on, say, renewable electricity, 225,000 Vermont homes would have gone green. Since we have only 240,000 households, that's pretty good."

"We have one question for the political leaders of the world," Kumi Naidoo, the international executive director of Greenpeace International, said at the huge climate rally held in Copenhagen halfway through the summit. "If you can find not millions, not billions, but trillions of dollars to bail out the banks, the bankers, and their bonuses, how is it that you cannot find the money to bail out the planet, the poor, and our children?"

Learning to Be Good Ancestors

Copenhagen was a terrible disappointment, but giving up is not an option. Despite the poor outcome, there is still time, as I write these words in early 2010, to pull victory from the jaws of defeat. The goal of the summit was to reach an agreement to take effect in 2012, when key provisions of the Kyoto Protocol expire; that timetable might still be met if governments make sufficient progress in the follow-up meetings scheduled for later in 2010 and 2011.

Nor was Copenhagen without signs of hope. One potential game changer was the emergence of a surprisingly muscular mass movement on behalf of climate action. Activists throughout the world have been calling for climate action ever since the UN Earth Summit in 1992, but never has civil society been half as visible or influential as it was in Copenhagen. Diverse, global, youthful, and unafraid to demand the supposedly impossible, the new climate movement is a force that governments, corporations, and other powerful institutions seem destined to reckon with for years to come. Of course the movement did not achieve all it wanted in Copenhagen — mass movements rarely succeed right away. But it did manage to put a key demand — reducing the amount of carbon dioxide in the atmosphere to 350 ppm — squarely on the public agenda. By the summit's final day, more than one hundred governments, representing more than half the nations on earth, had endorsed the 350 ppm target and its corollary of limiting temperature rise to 1.5°C above preindustrial levels. "The idea," President Mohamed Nasheed of Maldives told me, "is that people will agree not to murder others. Anything above 1.5°C, and we [in Maldives] have had it."

The test in the months and years to come is whether this movement can do what one of the ubiquitous black and yellow placards at the Copenhagen climate rally demanded: "Change the Politics, Not the Climate." More than anything, that test will determine whether the U-turn in behavior I've described in the past few pages can actually come to pass.

If you ask me whether I think this will happen, whether Green Apollo programs will be initiated in the United States and throughout the world, I can only paraphrase president-elect Obama: Yes we can, but only if we hold fast to unyielding hope. Make no mistake: hope is not merely a passive faith that things will turn out all right in the end. Hope is a verb, a choice, a commitment to fight for a better world no matter how long the odds appear. And I believe there are genuine reasons for hope.

First of all, the U-turn has already begun. Many communities and countries have made real progress in recent years in transforming popular consciousness and social behavior in more climate-friendly directions. In the United States, the federal government passed an economic stimulus package in 2009 that was essentially a scaled-down, domestic

version of the Green Apollo project; what's needed is to massively expand this initiative and extend it overseas.

Second, this U-turn could easily accelerate in the months and years
ahead because no one supports it more than the young. In the United
States, climate change has become the hottest issue on college and university campuses. By 2009, student pressure had led more than 650 college and university administrations to pledge to eliminate—not reduce,
eliminate—carbon pollution on their campuses. Many corporations are
also promising better environmental behavior, in part because students
are telling their recruiters that a company's greenness matters to where
they will choose to work after graduation.

The economy, too, is changing. For the first time, the world in 2008
invested more money in wind, solar, and other renewable power sources
than in fossil fuel or nuclear-sourced electricity. In the United States,
greenhouse gas emissions actually began declining. Lester Brown has
pointed out that between 2007 and 2009 U.S. emissions fell by 9 percent. Brown acknowledged that the economic recession was responsible
for much of this decline, but he argued that more environmentally significant factors also played a key role. U.S. companies and consumers
were using energy more efficiently, while coal-fired power plants were
being replaced by natural gas, wind, solar, and geothermal. Additional
government policy changes, including increasing the efficiency of autos
and appliances, suggest that these shifts could gain speed in the coming
years. Meanwhile, the U.S. government, the nation's largest single consumer of energy, has pledged to reduce its use of vehicles by 30 percent
by 2020 and to recycle 50 percent of its waste by 2015. "We do not yet
know how much we can cut carbon emissions because we are just beginning to make a serious effort," Brown concluded. "Whether we can
move fast enough to avoid catastrophic climate change remains to be
seen."

That's the big question: We've begun the journey toward more sustainable living, but can we accelerate our pace enough to avoid disaster?
Again, I believe we can, but only if we bring to bear the resources of
government and society as a whole through programs like Green Apollo
and activism like the climate movement that came of age in Copenhagen. Fortunately, Green Apollo programs make not only environmen-

tal but also economic and political sense. Because they would create jobs, spur technological innovation, and open vast opportunities for business, such programs are in the majority of people's self-interest and thus should attract considerable political support. But Green Apollo really would amount to almost a revolution in how politics is practiced, both in Washington and in capitals around the world. It would require fundamental changes in where government money goes, including taking billions of dollars away from some of the most powerful interests in the world, above all the oil and coal industries. Such fundamental change will happen only if government officials are pushed by intense, sustained public pressure. As Obama himself explained while running for president, "Change does not happen from the top down. It happens from the bottom up . . . [People] arguing, agitating, mobilizing, and ultimately forcing elected officials to be accountable. . . . That's how we're going to bring about change."

What stronger incentive do we need than the terrible oil spill that ravaged the Gulf of Mexico in 2010, just as this book was going to press? As I write these words, a live video feed shows the oil continuing to gush from BP's well some eighteen thousand feet below sea level. BP is plainly guilty of one of the great environmental crimes in history, but all of us who drive cars, ride in airplanes, use plastic, or otherwise consume petroleum products share the blame. The world's insatiable demand for oil is the reason BP was drilling at such tremendous depths in the first place. Extracting oil under those conditions is inherently risky: prepare for more disasters if we as a civilization don't leave oil behind soon. Since the earth seems to be reaching the point of peak oil, all future production carries a risk of similar catastrophes. Beyond the immediate necessity of plugging the leaking well, the real solution to the BP oil disaster is obvious: we must break our addiction to oil and instead embrace a future of clean energy and green jobs.

And a final paradox: we must act immediately even as we take the long view. Or, as Kevin Danaher, the cofounder and president of Global Exchange, a San Francisco–based NGO, put it, "We must learn to be good ancestors." Danaher, one of the world's leading activists working on creating local green economies, made that remark during a speech to a Green Festival conference in Italy in 2009. Visiting Florence, he had been struck by the beauty and craftsmanship embodied in the city's main cathedral, the incomparable black-and-white-stoned Duomo of Santa Maria del Fiore. The dedication and long-term vision required to erect such a structure during the Middle Ages was comparable, he told me, to what he and other champions of the emerging green economy must practice. "We are responsible for laying the foundations that future generations will build on," he said, "somewhat like the masons who laid the foundation layers of the European cathedrals that took several centuries to complete. They knew they would not live to see the final product of their work, but they also knew they needed to do very solid, precise work because of all the weight that was going to be placed on top of their work."

Being a good ancestor, said Danaher, means getting involved in all aspects of building a greener world: political engagement, grassroots economics, personal change. I would add that it also means starting right away. We don't know everything necessary to avoid the unmanageable and manage the unavoidable of climate change, but we don't have to. As Ron Sims commented about his own efforts in King County, our job is to begin, do the best we can, and trust others to carry on after our work is done. This was the guiding principle of the Renaissance geniuses who designed the Duomo, Danaher pointed out. They deliberately built the cathedral with a hole in the ceiling, awaiting the construction of a dome that was not yet technologically feasible. "The confidence of the Renaissance era was so great that they knew someone would come up with a way to engineer the dome, and the architect Filippo Brunelleschi did it," marveled Danaher, who added, "[R]egarding our environmental situation on this little blue marble, I believe a certain percentage of humanity will survive the coming collapse, and it will be the local, sustainable green economy that will be the base of that survival. If we can get the foundations [of that economy] right, future generations will figure out how to put the dome in place."

Epilogue: Chiara in the Year 2020

FRODO: I wish that none of this had ever happened.

GANDOLF: Of course you do. But that is not for you to decide. All you have to do is decide what to do with the time you have been given.

—J.R.R. TOLKIEN, *The Lord of the Rings*

Dear Chiara,

By the time you read this, you will be very grown up. I hope I'm there to see it. In any case, I've arranged for you to receive this letter on your fifteenth birthday, along with a copy of this book, which I had specially bound for you and this special day.

Since you're turning fifteen today, you'll be reading these words in the year 2020, a cardinal date for the challenge described in this book. According to the scientists I interviewed, many, many things have to happen by 2020 if this planet is to remain a livable place. But whether they are happening or not, today will still be your birthday, so first let me say, Happy birthday, Chiara! I hope you're having a spectacular day and the coming year brings lots of good things for you. Being your dad has been the great joy of my life.

I wonder what kind of birthday party you'll be having today. I can tell you that from the time you were very small, birthday parties have been pretty much your favorite thing in life. I'll never forget the day you turned two. You were so determined not to miss an instant of the celebration! An hour before the party began, there you were, already wearing your favorite party dress, sitting at your little table by the window, hands folded in front of you, eagerly waiting for the guests to arrive.

A few weeks ago, we celebrated your fifth birthday. This year, you decided to make it a tea party and invite only girls. (Luckily, you made an exception in my case.) The guest of honor was Sleeping Beauty. When she came traipsing through the garden, carrying a basket filled with treats and, it turned out, some magic tricks, you were incandescent with joy.

Do you remember any of this, now that you're fifteen?

And do you still believe in fairy tales, I hope? When I was fifteen, *The Hobbit* and especially *The Lord of the Rings* meant the world to me. I'm not sure why I loved them so much. Maybe because they taught me something I relearned when I had you as my daughter: there is magic in this world. Amazing things can happen, if you believe and do your part to make them happen. Trust in that magic, Chiara, share it with others, and it will take you far.

Now about this book: I put this copy aside for you because, well, probably everyone will have forgotten about it ten years from now. That's the way it is with most books. But I wanted you to know, now that you're old enough to understand what it means, that I wrote this book for you. Not only you, but for you first of all. I hope it made a difference.

I don't want to get too serious here. Hey, maybe those knuckleheads will be right and this whole global warming thing will turn out to be a bunch of nonsense. Wouldn't your father look silly then? Fox News, vindicated again! But you know what? I would gladly look the fool if it meant you didn't have to face—if we all didn't have to face—what virtually every scientist says is happening on this planet.

Now that you're fifteen, Chiara, you will be making more and bigger decisions for yourself. One question I often thought about while writing this book was where you should live in the future in order to stay safe. Soon, that decision will be yours to make. Choose carefully. If it were me, I'd look for a place that has a secure water supply, a capable government, and a vibrant community—a place where people know how to work with their hands, where they look out for one another and practice the Golden Rule. That's going to be your surest protection if things get difficult in the years ahead.

I don't think there's much more I can tell you, Chiara. It'd be kind of silly for me to try. You'll know so much in 2020 that I don't know today. By then, scientists should be able to forecast with much greater precision what kind of impacts the world will experience from climate change, how soon, and where. It will also be pretty clear how well societies are preparing themselves against these impacts. Above all, in 2020 you'll know whether our country and the world as a whole are making—not just promising, but making—the dramatic reductions in greenhouse gas emissions that are needed for your generation to avoid the worst scenarios of climate change.

From where I sit, all that lies in the future. At this point, my precious, beautiful daughter, all that's clear is that our civilization is entering a storm. There is no way around it; we have to go through it. We have to be brave, resourceful, and never give up. I would give my life to see you safe on the other side.

With more love than I can say,
Daddy

Acknowledgments

A journalist is only as good as his or her sources. So I want to begin these acknowledgments by expressing my deep, enduring gratitude to the many people who took time out of their busy schedules to answer my questions about climate change and how our civilization can best confront it. Most of these individuals are named in the text. But I want to single out the ones who went an extra step and reviewed parts of the manuscript to alert me to any errors of fact, tone, or interpretation (though I stress that they bear no responsibility for the final text): Robert Bea, Laurens Bouwer, Ian Burton, Kim Cahill, Kevin Danaher, Mark Davis, Peter Gleick, Mark Goldthorpe, David Graves, James Hansen, Jerry Hatfield, Oliver Houck, Veronika Huber, Saleemul Huq, Jiang Gaoming, Sadhu Johnston, Gregory Jones, David King, Richard Klein, Aalt Leusink, Lin Erda, Jim Lopez, Mark Lynas, Ma Shiming, Michael Oppenheimer, David Pierce, Chris Reij, Cynthia Rosenzweig, Hans Joachim Schellnhuber, Sara Scherr, David Shearer, Will Travis, Ivor van Heerden, Pier Vellinga, Jim Verhey, Chris West, Robert Wilkinson, and Don Wuebbles. A number of friends were also generous enough to read parts of the manuscript and give me their comments; my warmest thanks to Denny May, Jonathan King, Toni Whiteman, Cliff Cunnington, Sasha West, and Ross Gelbspan.

Scarcely less important than sources are the people who connect a journalist to the sources. Here, I wish to thank the individuals who facilitated my travels, suggested people and places to see, or who served as interpreters: in Bangladesh, Atiq Rahman, Saleemul Huq, and Mozaharul Alam of the Bangladesh Centre for Advanced Studies and Isabelle Lemaire of the International Institute for Environment and Development; in Burkina Faso and Mali, Chris Reij; in Japan, Charles Scawthorn and Aaron Isgar; in China, Yixiu Wu, Liang Yan, and Jasper Becker; in New Orleans and Florida, Mark Schleifstein and Darryl Malek-Wiley; in the Netherlands, Madelene Helmer and Henk van Schaik.

In an age when serious journalism is an endangered species, I also benefited from magazine editors who were willing to spend the money and time necessary to develop this story. I especially thank Graydon Carter and Michael Hogan at *Vanity Fair,* who in the wake of Hurricane Katrina agreed to back the investigation that became the cover story of *Vanity Fair*'s first "green" issue and then gave rise to this book. *Vanity Fair*'s Cullen Murphy later read the entire manuscript and offered valuable encouragement and editorial advice. My longtime colleagues at the *Nation* never wavered in their support or interest; I thank in particular Katrina vanden Heuvel, Karen Rothmyer, and Betsy Reed. At *Time,* I have enjoyed working with Charles Alexander, Jyotti Thottam, and Jeffrey Kluger. At *L'espresso,* Antonio Carlucci has been an enthusiastic and attentive editor (thanks also for the introduction from Federico Rampini). I have been fortunate to collaborate with radio colleagues as well: thanks to Elizabeth Tucker at American Public Media's *Marketplace* and Peter Thompson at Public Radio International's *The World.*

I would not have been able to complete this book without the financial and moral support provided by the Open Society Institute. I received two grants from OSI: the first, in the wake of Hurricane Katrina, enabled me to make my first post-hurricane visit to New Orleans and to Tampa Bay; the second, a year spent as an OSI Fellow, financed my research trips to China, west Africa, and Copenhagen, covered my living expenses while I finished writing the book, and paid for the professional fact checking of the text done by journalist Tay Wiles. I cannot thank OSI enough. My gratitude goes in particular to George Soros, Leonard Bernado, Nancy Youman, Bipasha Ray, Lisena Desantis, Bronwen Manby, and above all Steve Hubbell, who first urged me to apply for the OSI fellowship. Special thanks as well to Michael Vachon, who, along with my old pal David Fenton, helped decide the book's title.

I am also grateful to author Lutz Kleveman and the Drager Foundation for bringing me to Germany to participate in the 2006 Ankelohe Conversation about the intersection between climate change and peak oil.

I am proud to have this book published by Houghton Mifflin Harcourt. Houghton Mifflin has been America's premier publisher of environmental books dating back to Henry David Thoreau and up through Rachel Carson and Al Gore. I am especially grateful to Anton Mueller, who first commissioned this book and believed in its premise and author from the start. Later, I benefited beyond words from the editorial guidance and moral support of George Hodgman, who tirelessly pushed me to make the text as accessible as possible and then enthusiastically championed the result. I am grateful as well to Barbara Wood, whose wordsmithing saved me from countless infelicitous phrasings.

As always, my agent, Ellen Levine, has been a vital partner, adviser, reader, and cheerleader. I hope she'll never retire.

Finally, I wish to thank my friends and family, whose support has been essential and whose love has been my rock. You know who you are.

Notes

This book is based on more than four years of original reporting and research, including travel to destinations across the United States and around the world. In these notes, I provide documentation for specific statements made in the text, as well as comments and suggestions for further reading. Please note that the great majority of quotations included in this book are derived from interviews that I did myself; likewise, most descriptions of people and places are based on my eyewitness observations. To avoid redundancy, I usually do not provide individual references for each of these descriptions and quotations. Rather, I document only questions of fact that come from other sources. Thus the reader can assume that if no reference is provided for a particular quote or description, that item came from my own reporting.

These notes were assembled with the assistance and expertise of journalist Taylor Wiles, who fact-checked the entire manuscript, helped track down elusive documentation, and corrected errors. In addition, I sent selected passages of the manuscript to outside experts, including some featured in the text, so they could offer comments and corrections, a process that saved me from making a number of inadvertent mistakes and imprecise observations.

I could not have written this book without relying on the voluminous scientific research that has been done on global warming and climate change and, equally important, the efforts of scientifically literate experts to explain those findings in ways that a non-scientist such as myself can understand. A foundation source is the Intergovernmental Panel on Climate Change. The IPCC has been criticized over the years, both by deniers of climate change who focus on a handful of errors in thousands of pages of text to try to discredit the entirety of climate science and, on the other side, by scientists and advocates who complain that the IPCC's procedures (including the control that governments exercise over the executive summaries of IPCC assessments) make its reports overly conservative and dated. Nevertheless, the IPCC's reports, especially its four *Assessment Reports* (published in 1990, 1995, 2001, and 2007), are necessary (if often dry and technical) reading for any student of climate change. Beyond the IPCC, Mark Lynas, a science writer based in Oxford, has written one of the essential books on the subject, *Six Degrees: Our Future on a Hotter Planet* (London: Harper Collins, 2007). Joe Romm, a physicist and former assistant secretary of energy under President Bill Clinton, is also well worth reading. His blog, *Climate Progress,* provides comprehensive, timely analysis of the latest scientific developments (with links to the studies under discussion), all explained in an accessible style that, though unabashedly opinionated, arms the reader with "the power which knowledge gives," in Madison's imperishable phrase. Unlike Romm and Lynas, Bill McKibben is not a trained scientist, but he is unexcelled at explaining the larger implications of climate science for our civilization; dating back to *The End of Nature* in 1989, the first major popular book on climate change, through his current journalism and his 2010 book *Eaarth: Making a Life on a Tough New Planet* (New York: Times Books, 2010), McKibben is the dean of climate writers, the one in whose footsteps the rest of us follow. Among daily journalists, I find the work of Richard Black of the BBC and Fiona Harvey of the *Financial Times* indispensable.

One final caution: the science and politics of climate change have evolved very rapidly in recent years, so some of the documentation found here may have become dated by the time you read these words. I urge readers to continue to explore the preceding sources of information, as well as my own website, http://www.markhertsgaard.com, where I will post updates as warranted.

Prologue: Growing Up Under Global Warming

Breaking my own rule here at the beginning, let me underline that the quote from Martin Parry, like most other quotes in this book, comes from an author's interview. The descriptions of the fate of children in Russia, Africa, and China are based on travels I made in the 1990s and are reported in more detail in my book *Earth Odyssey: Around the World in Search of Our Environmental Future* (New York: Broadway Books, 1998). David King's "most severe problem" quote comes from his article "Climate Change Science: Adapt, Mitigate, or Ignore?" in *Science* 303, no. 5655 (January 9, 2004): 176–77. The biographical material about him was mostly provided in three separate author's interviews. James Hansen's 1988 Senate testimony was based on the paper by J. Hansen, I. Fung, A. Lacis, D. Rind, S. Lebedeff, R. Ruedy, G. Russell, and P. Stone, "1988: Global Climate Changes as Forecast by Goddard Institute for Space Studies Three-Dimensional Model," *Journal of Geophysical Research* 93, 9341–64, doi:10.1029/88JD00231. The passage outlining the differences between *global warming* and *climate change* is based on my own reading of the scientific literature and interviews with numerous climate scientists, including King, Hansen, and Michael Oppenheimer of Princeton. By separating global warming from climate change, I am making explicit the crucial role played by what scientists call the sensitivity of the climate system: that is, how much temperature rise (i.e., global warming) is required before attributable impacts (i.e., climate change) are triggered?

The disinformation campaign involving Frederick Seitz and the Global Climate Coalition is detailed in Chapter 10; hence, documentation is provided there. The memo unearthed by Gelbspan is described in his book *The Heat Is On: The High Stakes Battle over Earth's Threatened Climate* (Reading, MA: Addison-Wesley, 1997). See also his website, http://www.theheatison.org. Via his then-aide Marc Morano, Senator James Inhofe repeated his "greatest hoax" quote to me during an e-mail interview conducted for my article "While Washington Slept," *Vanity Fair,* May 2006.

The groundbreaking "climate signal" study of the 2003 heat wave was conducted by scientists Peter A. Stott of the University of Reading and UK Met Office and D. A. Stone and M. R. Allen, both of Oxford University, and published as "Human Contribution to the European Heat Wave of 2003" in *Nature* 432 (December 2, 2004): 610–14. The accumulation of corpses outside of morgues in Paris in 2003 was reported by the *New York Times* on August 15, 2003. The 31,000 mortality figure King gave for the 2003 heat wave was based on estimates provided by the governments of the countries affected. Researchers for the European Union later concluded that this figure was a gross underestimate; see Chapter 3 for details.

The concentration and characteristics of carbon dioxide and other greenhouse gases are most authoritatively described in Contribution of Working Group I to the *Fourth Assessment Report* of the Intergovernmental Panel on Climate Change, edited by S. Solomon, D. Qin, M. Manning, Z. Chen, M. Marquis, K. B. Averyt, M. Tignor, and H. L. Miller (Cambridge and New York: Cambridge University Press, 2007). Hereafter in these notes, the *Fourth Assessment Report* will be abbreviated as *4AR*. Leggett coined the phrase "the carbon club lobby" in his book *The Carbon War: Global Warming and the End of the Oil Era* (London: Routledge, 2001). A former geologist for the oil industry turned chief scientist for the NGO Greenpeace International, Leggett began attending the international negotiations on climate change in 1990. At the time there were relatively few people involved, giving Leggett an insider's vantage point on the deliberations. His book reveals how overt, extensive, and powerful were the steps taken by fossil fuel interests to block action. The Global Climate Coalition's rejection and censoring of its own scientists' conclusions were reported in the *New York Times* on April 23, 2009; see http://www.nytimes.com/2009/04/24/science/earth/24deny.html.

Chapter I: Living Through the Storm

The "gambling the planet" quote comes from *The Global Deal: Climate Change and the Creation of a New Era of Progress and Prosperity* by Nicholas Stern (New York: Public Affairs, 2009). Bruno Bettelheim's comment about fairy tales is found in *The Uses of Enchantment: The Meaning and Importance of Fairy Tales* (New York: Random House,

1976). Lisa Bennett's quote came from her article "The Hot Spot," originally published in *Greater Good Magazine* (Fall 2008) and later reprinted in *The Compassionate Instinct: The Science of Human Goodness*, edited by D. Keltner, J. March, and J. A. Smith (New York: W. W. Norton, 2010). G. K. Chesterton's quote is from his book *Tremendous Trifles* (New York: Dodd, Mead and Company, 1909).

The BP example was described by Sir Charles Nicholson, BP's group adviser on the environment, in an article published February 25, 2004, in the Middle East business newsletter *AMEinfo.com*. The apartment retrofitting undertaken by the German government was described to me in interviews with various German political leaders and public officials, especially Reinhard Bütikofer, then the chair of the German Green Party. See my article "Green Power," *Nation*, January 30, 2006. California's achievements on energy efficiency are best described in Part 4 of Joe Romm's five-part series on energy efficiency, published on his blog, *Climate Progress*. See http://climateprogress.org/2008/07/30/energy-efficiency-part-4-how-does-california-do-it-so-consistently-and-cost-effectively/. The HSBC Global Research study was summarized in the business newsletter *Environmental Leader*, September 22, 2009.

Documentation for the 12.5 million acres of land rehabilitated in the Sahel is provided in Chapter 8, where the story is told in full. James Schlesinger's "peak-ists" statement was made on September 17, 2007, during his keynote speech to a conference of the Association for the Study of Peak Oil and Gas and was reported in that day's issue of *Energy Bulletin*. Ron Oxburgh's quote came in an author's interview. Fatih Birol's quote came in an article he wrote for London's *Independent* newspaper, March 2, 2008, available at http://www.independent.co.uk/news/business/comment/outside-view-we-cant-cling-to-crude-we-should-leave-oil-before-it-leaves-us-790178.html.

The *New York Times* story about NASA scientist James Hansen's testimony is available at http://www.nytimes.com/1988/06/24/us/global-warming-has-begun-expert-tells-senate.html. Nicholas Stern's "greatest and widest-ranging market failure" quote is found in the executive summary of *The Economics of Climate Change: The Stern Review* (Cambridge University Press, 2007); see http://webarchive.nationalarchives.gov.uk/+/http://www.hm-treasury.gov.uk/d/Executive_Summary.pdf. The European Union's report on climate change's security implications, written by Javier Solana and Benita Ferrero-Waldner, was reported in the *Guardian*, March 10, 2008. The study by the American Geophysical Union, the American Meteorological Society, and other leading scientific authorities is available at http://www.ucar.edu/td/. The statement about biodiversity appeared in *Nature* on July 19, 2006: Michel Loreau, Alfred Oteng-Yeboah, M. T. K. Arroyo, D. Babin, R. Barbault, M. Donoghue, M. Gadgil, C. Häuser, C. Heip, A. Larigauderie, K. Ma, G. Mace, H. A. Mooney, C. Perrings, P. Raven, J. Sarukhan, P. Schei, R. J. Scholes, and R. T. Watson, "Diversity Without Representation," *Nature* 442 (July 20, 2006), 245–46 | doi:10.1038/442245a; published online July 19, 2006. Richard Louv's comments were delivered in a public talk he gave in Mill Valley, California, on May 1, 2008, which I attended. Hansen's "excuse" quote is found at http://www.columbia.edu/~jeh1/mailings/2008/20081023_Obstruction.pdf. I was part of the Chicago Humanities Festival panel on which George Woodwell appeared.

Chapter 2: Three Feet of Water

The IPCC's conclusion that sea levels will keep rising for centuries and oceans expanding for millennia is found in *4AR* (Working Group II). Lynas made his statement about "timing" at the Ankelohe Conversations Symposium on Climate Change and Peak Oil, held in Gut Ankelohe, Germany, May 2006, which I attended. The source for the 145 million people affected by three feet of sea level rise is the UN Environment Programme, drawing on a study by the Tyndall Centre for Climate Change Research in Norwich, England; see http://maps.grida.no/go/graphic/population-area-and-economy-affected-by-a-1-m-sea-level-rise-global-and-regional-estimates-based-on-. The references to Manila, Jakarta, and Dhaka were reported by Reuters on November 12, 2009. The estimate of $3 trillion of assets endangered by sea level rise are from *The Stern Review: The Economics of Climate Change;* see http://webarchive.nationalarchives.gov .uk/+/http://www.hm-treasury.gov.uk/independent_reviews/stern_review_economics_ climate_change/stern_review_report.cfm/. Note that Stern declared in 2009 that his review, published in 2006, had unintentionally "under-estimated" the problem. The list of sites at risk from three feet of sea level rise is based, in the cases of New York, London, Tokyo, and Shanghai, on the author's own on-site reporting, augmented by data on the cities' respective topographies. The impacts of sea level rise on the U.S. transportation system are described in *Potential Impacts of Climate Change on U.S. Transportation,* available at http://onlinepubs.trb.org/onlinepubs/sr/sr290.pdf. Lloyd's of London adviser David Smith's remarks on sea level rise can be viewed at http://www.lloyds.com/ News_Centre/360_risk_insight/Expert_opinion/Coastal_communities_and_climate_ change.htm. Among many other sources that find two meters of sea level rise by 2100 a credible scenario, see "Kinematic Constraints on Glacier Contributions to 21st Century Sea-Level Rise," by W. T. Pfeffer, J. T. Harper, and S. O'Neel, in *Science* 321, no. 5894 (September 5, 2008): 1340–43. The Pacific Institute study, "The Impacts of Sea-Level Rise on the California Coast," was published in May 2009 and is available at http://www.pacinst .org/reports/sea_level_rise/report.pdf. I heard Will Travis's presentation by attending the San Francisco Public Utilities Commission hearing, which was open to the public.

The 92 percent figure for Bangladesh is reported by the UNFCCC; see http://unfccc .int/resource/docs/napa/ban01.pdf. The effect of sea level rise on salinity and rice yields in southern Bangladesh is documented in *Investigating the Impact of Relative Sea-Level Rise on Coastal Communities and Their Livelihoods in Bangladesh,* a study funded by the British government's Department for Environment, Food and Rural Affairs and conducted by the Institute of Water Modeling of the Bangladesh Ministry of Water Resources, June 2007. The 41 percent malnutrition rate in Bangladesh refers to children under age five and is reported by the World Bank; see http://devdata.worldbank.org/ AAG/bgd_aag.pdf. Henry Chu's article appeared in the *Los Angeles Times* on February 21, 2007. Justin Huggler's article appeared in the *Independent* of London on February 19, 2007. The inundation effects of one and two meters of sea level rise are described in the Bangladesh government's *Climate Change Strategy and Action Plan 2008,* available at http://www.sdnbd.org/moef.pdf.

The effect of sea level rise on the frequency of 1-in-100-years floods in New York is described in the city government's climate action plan, PlaNYC, available at http://www .nyc.gov/html/planyc2030. Information on the 1938 hurricane that struck New York is available from the National Hurricane Center of the U.S. National Oceanic and Atmospheric Administration; see http://www.nhc.noaa.gov/HAW2/english/history.shtml#new. That New York is historically overdue for a hurricane has been noted by, among others, Jeffrey Schultz, a climatologist at the Northeast Regional Climate Center; see the New York 1 news report from 2006 at http://www.ny1.com/?SecID=1000&ArID=60604.

Chapter 3: My Daughter's Earth

John Holdren's quote was contained in the address he delivered as the president of the U.S. National Academy of Sciences in January 2008, reprinted in *Science,* January 25, 2008. The Podesta war game, which I attended, was cosponsored by the Center for a New American Security; see the center's website for full details: http://www.cnas.org/ node/956.

The study of the 2003 heat wave done for the European Union was "Death Toll Exceeded 70,000 in Europe in the Summer of 2003," by Jean-Marie Robine et al., *Les Comptes Rendus/Série Biologies* 331 (2008): 171–78. The information on present and projected summer heat in New York City comes from the city government's climate action plan, PlaNYC, cited in the notes for Chapter 2. The definitive source on the Chicago heat wave of 1995 is *Heat Wave: A Social Autopsy of Disaster in Chicago* by Eric Klinenberg (Chicago: University of Chicago Press, 2002). The information on St. Louis and Ohio comes from a report by the Union of Concerned Scientists, *Confronting Climate Change in the U.S. Midwest,* which can be found at http://www.ucsusa.org/global_ warming/science_and_impacts/impacts/climate-change-midwest.html. The effects on hydro and nuclear power plants are described in *Six Degrees* by Mark Lynas, cited in the introduction to the Notes, and in the *Guardian* of August 13, 2003. The effects of the 2006 heat wave in California were described in an author's interview by professor Robert Wilkinson of the University of California, Santa Barbara, who worked closely with electric utility companies while coordinating the California portion of the *First U.S. National Assessment* on climate impacts, which was published in 2000 and is available at http://www.globalchange.gov/publications/reports/scientific-assessments/first-national-assessment. Further information on the 2006 heat wave was compiled by the Northwest Power and Conservation Council: http://www.nwppc.org/energy/resource/ meetings/2006/08/2006%20Heat%20CAISO.pdf.

The IPCC's projection of increased severity of hurricanes is found in the *4AR*. The *Nature* study is "The Increasing Intensity of the Strongest Cyclones" by James B. Elsner, James P. Kossin, and Thomas H. Jagger, *Nature* 455 (September 4, 2008): 92–95. Munich Re's findings are available at http://www.munichre.com/publications/302-06295_en.pdf, page 37. John Holmes's comments about 2007 were made at a press conference described in a Reuters dispatch, January 30, 2008. His comments on 2008 were reported by

Agence France-Presse, June 17, 2009. Oxfam's concerns were reported in the *Guardian* of April 21, 2009. The original *Up in Smoke* report, as well as various sequels, are available from the New Economics Foundation: http://www.neweconomics.org/publications/smoke-reports-summary.

The *Lancet* article, by Anthony Costello, was published in volume 373, issue 9676 (May 16, 2009): 1669. The 150,000 annual death toll is found in the *4AR*, Working Group II, Section 8.4.1.1. The history and shortcomings of that estimate were described by Harvard's Paul Epstein in an author's interview. The discovery of malaria in Palau, New Guinea, was recounted by public health expert Kris Ebi in an author's interview. The increased incidence of asthma is projected by the WHO in its "Climate Change and Health" web page: http://www.who.int/mediacentre/factsheets/fs266/en/index.html. The possible resurgence of dengue fever is discussed in "Dengue and Hemorrhagic Fever: A Potential Threat to Public Health in the United States" by David M. Morens, MD, and Antony S. Fauci, MD, *Journal of the American Medical Association* 299, no. 2 (2008): 214–16.

The *4AR*'s projection of drought in the U.S. Southwest is from "Model Projections of an Imminent Transition to a More Arid Climate in Southwestern North America," by Richard Seager et al., *Science* 316 (May 25, 2007), pp. 1181–84. The projection for Lake Mead is found in "When Will Lake Mead Go Dry?" by T. P. Barnett and D. W. Pierce, Water Resources Research, doi:10.1029/2007WR006704, in press; a nonscholarly summary of the report is available from the Scripps Institution of Oceanography, where both Barnett and Pierce work: http://scrippsnews.ucsd.edu/Releases/?releaseID=876. I also later interviewed Pierce about the study. The figures about snowpack and water supply in California come from interviews with state water experts; since this subject is discussed at length in Chapter 9, documentation is provided there. The parallels to the water situation in the Pacific Northwest are based on interviews with experts there and especially on the scholarly work of the University of Washington's Climate Impacts Group; since that subject is explored in Chapter 4, check there for documentation. The data and quotes about the Himalayan snowpack were reported by Orville Schell in "The Message from the Glaciers," *New York Review of Books*, May 27, 2010. Lynas's reporting on the Rwenzori is found in *Six Degrees*. The situation in South America is described by Lynas, but see also "Deglaciation in the Andean Region" by James Painter, a UNDP study for the Human Development Report 2007/2008, available at http://78.136.31.142/en/reports/global/hdr2007-2008/papers/Painter_James.pdf. The "elephant in the room" quote is from *The Great Warming: Climate Change and the Rise and Fall of Civilizations* by Brian Fagan (New York: Bloomsbury Press, 2008). The increase in water-stressed people is projected in the UNDP's Human Development Report 2006, *Beyond Scarcity: Power, Poverty, and the Global Water Crisis*, available at http://hdr.undp.org/en/reports/global/hdr2006/.

David Lobell's quote came in an author's interview. Lynas's prairies projection is from *Six Degrees*. China's projection of a 37 percent decline in crop yield appeared in the government's *National Climate Change Programme*, which was published in June 2007 and is available at http://en.ndrc.gov.cn/newsrelease/P020070604561191006823.pdf. The projections for South Asia and Africa are found in the *4AR*, Working Group I, Section 11.2 Africa.

Lynas's description of 2003 wildfires is from *Six Degrees*. The firefighters' statement, known as the San Diego Declaration, is at http://www.fireecology.net/Climate-Change-and-Fire-Management/. The Harvard study, by Jennifer Logan, was published in the *Journal of Geophysical Research* in June 2009; a useful summary is here: http://harvardscience.harvard.edu/engineering-technology/articles/scientists-expect-wildfires-increase-climate-warms-coming-decades.

The fate of coral reefs was projected by Ken Caldeira and colleagues in "Coral Reefs Under Rapid Climate Change and Ocean Acidification" by O. Hoegh-Guldberg, P. J. Mumby, A. J. Hooten, R. S. Steneck, P. Greenfield, E. Gomez, C. D. Harvell, P. F. Sale, A. J. Edwards, K. Caldeira, N. Knowlton, C. M. Eakin, R. Iglesias-Prieto, N. Muthiga, R. H. Bradbury, A. Dubi, and M. E. Hatziolos, *Science* 318 (December 14, 2007): 1737–42. The U.S. Geological Survey's projection about polar bears, "Predicting the Future Distribution of Polar Bear Habitat in the Polar Basin from Resource Selection Functions Applied to 21st Century General Circulation Model Projections of Sea Ice," is available at http://www.usgs.gov/newsroom/special/polar_bears/docs/USGS_PolarBear_Durner_Habitat_lowres.pdf. The information on acidification of the oceans is drawn from Bill McKibben's book *Eaarth* (cited in the introduction to the Notes), pages 9–10. For a fuller treatment of the topic, see Elizabeth Kolbert's article "The Darkening Sea," in the *New Yorker*, November 20, 2006. The information on Arctic summer ice melt and the quote from Jay Zwally were reported by the Associated Press December 12, 2007. The economic importance of healthy ecosystems is documented and explained in the interim report of the European Commission's *The Economics of Ecosystems and Biodiversity*, released in 2008 and available at http://ec.europa.eu/environment/nature/biodiversity/economics/pdf/teeb_report.pdf. E. O. Wilson's quote about ants is found in his foreword to *Sustaining Life: How Human Health Depends on Biodiversity*, edited by Eric Chivian and Aaron Bernstein (New York: Oxford University Press, 2008).

The role of the Bush administration in killing the U.S. National Assessment has been described by the whistleblower Rick Piltz, who after leaving the government established the organization Climate Science Watch; for information, visit http://www.climatesciencewatch.org/. As evidenced by John Marburger's quote, the administration also invoked the adaptation option as an argument against pursuing mitigation, though it was by no means the first to do so. One of the most important earlier voices was that of economist William Nordhaus, whose many studies in the early 1990s argued that it would be cheaper for society to allow global warming to occur (if the scientific warnings in fact turned out to be accurate) and then use the wealth generated by unencumbered fossil fuel burning to adapt to impacts decades in the future. The study was frequently cited in media and congressional discussions of climate policy in the 1990s; see a listing of his many articles at http://nordhaus.econ.yale.edu/cv_current.htm. See also Bill McKibben's observations in *Deep Economy: The Wealth of Communities and the Durable*

Future (New York: Henry Holt, 2007), page 25. Gore described his evolving views on adaptation in an article in *Economist* of September 11, 2008.

The carbon content of the permafrost and quote from Canadell are found at http://www.csiro.au/news/PermafrostCarbon.html. The strength of Hurricane Katrina, which varied from Category 1 at its outset to Category 5 in the Gulf of Mexico and Category 3 at landfall, is documented by the U.S. government's National Hurricane Center. Schellnhuber's 2008 study is T. M. Lenton, H. Held, E. Kriegler, J. W. Hall, W. Lucht, S. Rahmstorf, and H. J. Schellnhuber, "Tipping Elements in the Earth's Climate System," *Proceedings of the National Academy of Sciences,* Online Early Edition, February 4, 2008. The countries endorsing a 1.5°C temperature limit are listed at http://www.350.org. The 0.18°C of temperature rise per decade was reported in January 2010 by the U.S. National Oceanic and Atmospheric Administration (NOAA) in its monthly "State of the Climate Global Analysis": http://www.ncdc.noaa.gov/sotc/?report=global.

Robert Watson and Mohamed El-Ashry's article on methane appeared in the *Wall Street Journal* of December 28, 2009. Ecoagriculture's article appeared as Chapter 3 of *State of the World 2009: Into a Warming World* by the Worldwatch Institute (New York: W. W. Norton, 2009). The most up-to-date discussion of geoengineering for the nonspecialist reader is supplied by Jeff Goodell in *How to Cool the Planet: Geoengineering and the Audacious Quest to Fix Earth's Climate* (New York: Houghton Mifflin Harcourt, 2010). John Holdren's statement, uttered in May 2009, was reported by Joe Romm in a January 27, 2010, post at his *Climate Progress* blog; a post on January 6, 2010, describes the latest research on white roofs and pavements.

Chapter 4: Ask the Climate Question

The Wasco tribe's folktale, and the other tales recounted later in this chapter, are found in *Legends of the Pacific Northwest* by Ella E. Clark (Berkeley: University of California Press, 1953). The "Mr. Salmon" nickname for Ron Sims was shared with me by Dennis Hayes, the president of the Bullitt Foundation in Seattle and an occasional adviser to Sims. The passage on the Greenbridge and Overlake Station housing developments is based on my interview with Stephen Norman, supplemented by printed materials supplied by the Authority's website: http://www.kcha.org.

The *Seattle Times* editorial was published on September 7, 1988. S. Fred Singer's receipt of funding from fossil fuel interests was documented in *The Heat Is On,* cited previously. The biographical information on Sims was provided by him in an author's interview. The reports of the Climate Impacts Group are available at http://cses.washington.edu/cig/. The emissions reduction efforts of cities under the Cities for Climate Protection campaign are described on the website of ICLEI—Local Governments for Sustainability: http://www.iclei.org/index.php?id=10829. The actions of King County and the city of Seattle are described on their respective websites, supplemented by information from Sims in an author's interview. Information on the October 2005 conference and guidebook is available at http://www.kingcounty.gov/exec/globalwarming/

environmental/2005-climate-change-conference.aspx. Background information on the Brightwater water treatment facility is available at http://www.kingcounty.gov/environ ment/wtd/Construction/North/Brightwater.aspx. The *Ask the Climate Question* report is at http://www.ccap.org/docs/resources/674/Urban_Climate_Adaptation-FINAL_CCAP %206-9-09.pdf. The quotes from Steve Winkelman and Elizabeth Willmott came in author's interviews.

Chicago's climate action plan is available at http://www.chicagoclimateaction.org/. The Clinton Climate Initiative's work is described at http://clintonfoundation.org/what-we-do/clinton-climate-initiative/our-approach/cities. The activities of the Partnership for Sustainable Communities are described at http://www.epa.gov/smartgrowth/partnership/index.html.

I witnessed Mayor Bloomberg's speech on Earth Day 2007, which is available, with supporting information about PlaNYC, at http://www.nyc.gov/html/planyc2030/html/home/home.shtml. The work of the city's Panel on Climate Change is at http://www .nyc.gov/html/om/pdf/2009/NPCC_CRI.pdf. Oppenheimer's vision of the siting for such a seawall was communicated in an author's interview, as were Thomas Frieden's views on health impacts. The New York legislature's rejection of congestion pricing, voted on April 7, 2008, was reported in contemporaneous press accounts; its rejection of additional energy efficiency funding was described by Aggarwala in an author's interview.

The information and resources of NOAA's Climate Services program are available at http://www.climate.gov/#climateWatch.

Chapter 5: The Two-Hundred-Year Plan

The "no reason to panic" quote, as well as additional facts and insights contained later in Chapter 5, comes from the 2008 report of the Sustainable Coastal Development Commission, *Working Together with Water: A Living Land Builds for Its Future*, available in English at http://www.deltacommissie.com/doc/deltareport_summary.pdf. The elevation of Schiphol airport was mentioned by Aalt Leusink in an author's interview and is confirmed at the airport website: http://www.schiphol.nl/SchipholGroup/Company1/Profile/Activities/AmsterdamAirportSchiphol.htm. For much of the background information in this chapter about water in the Netherlands, including the "drain of Europe" phrase, I relied on the many publications offered by the Ministry of Transport, Public Works and Water Management, including *Water in the Netherlands:* http://www.safecoast.org/editor/databank/File/Water%20in%20NL%202004-2005,%20 facts%20and%20figures.pdf. Aalt Leusink's comments about the European perspective on the Netherlands and the country's two-hundred-year plan, and his remaining comments in the chapter, came from author's interviews; the same is true of Pier Vellinga's quotes. The 1275 date for the first dam across the Amstel River is cited by the Amsterdam Historical Museum: http://www.channels.nl/amsterdam/historic.html.

Useful background on the Dutch adaptation plan is found in an article Leusink co-authored, with Michiel van Drunen and Ralph Lasage, "Towards a Climate-Proof Netherlands," which appears in the book *Water Management in 2020 and Beyond*, edited by

Asit K. Biswas, Cecilia Tortajada, and Rafael Izquierdo (Berlin: Springer Verlag, 2009). Another essential resource is the Knowledge for Climate initiative directed by Pier Vellinga: http://knowledgeforclimate.climateresearchnetherlands.nl/nl/25222734-Home .html. The deaths and damages attributed to the 1953 flood, as well as the effort to recover and construct the Delta Works, were described in author's interviews with Leusink, Vellinga, and Roelvink and are detailed in the Delta Works's own publication, *Water, Nature, People, Technology,* available at http://www.deltaworks.org/downloads/ summaries/PDF/english_pdf_deltaworks.org.pdf. Pavel Kabat's study of adaptation in Zeeland was delivered at the Salzburg Global Seminars and is available at http://www .salzburgglobal.org/mediafiles/PRES1967.pdf. See also David Roberts's report in *Grist,* July 11, 2008. The Green Plan of the Netherlands is best described (in English) in *Green Plans: A Blueprint for a Sustainable Earth* by Huey D. Johnson (Omaha: University of Nebraska Press, 2008).

Chapter 6: Do You Know What It Means to Miss New Orleans?

The descriptions of Hurricane Katrina in this chapter are based partly on author's interviews but also draw on the enormous amount of often excellent journalism, books, and scientific investigations that have been published about the storm and its aftermath. Among the essential sources are, first and foremost, *Path of Destruction: The Devastation of New Orleans and the Coming Age of Superstorms* (New York: Little, Brown, 2007), by John McQuaid and Mark Schleifstein, Pulitzer Prize–winning journalists at the New Orleans *Times-Picayune* newspaper. The staff of the *Times-Picayune* was heroic during and after Katrina, managing to publish every day, at least on the web, despite being forced to evacuate their offices; their reporting was and remains the best available chronicle of what happened when and why. The science of hurricanes and the failures of the Louisiana levee system—and how those failures could be corrected—are best described in *The Storm: What Went Wrong and Why During Hurricane Katrina—the Inside Story from One Louisiana Scientist* (New York: Viking, 2006) by Ivor van Heerden and Mike Bryan. The human consequences of Katrina are superbly rendered in *Breach of Faith: Hurricane Katrina and the Near Death of a Great American City* (New York: Random House, 2006) by Jed Horne, another reporter at the *Times-Picayune.* Also worth reading is *The Great Deluge: Hurricane Katrina, New Orleans, and the Mississippi Gulf Coast* (New York: William Morrow, 2006) by Douglas Brinkley. Three scientific reports are must reading: first, the Army Corps of Engineers' self-examination, *Performance Evaluation of the New Orleans and Southeast Louisiana Hurricane Protection System,* a nine-volume study which admitted that the New Orleans levee system was "a system in name only" and whose executive summary is available at http://www.nytimes.com/packages/ pdf/national/20060601_ARMYCORPS_SUMM.pdf; second, an investigation funded by the National Science Foundation, *Investigation of the Performance of the New Orleans Flood Protection Systems in Hurricane Katrina on August 29, 2005,* by Raymond Seed, Robert Bea, et al. (2006), available at http://www.ce.berkeley.edu/projects/neworleans/; finally, the state of Louisiana's investigation, *The Failure of the New Orleans Levee System*

During Hurricane Katrina, by Ivor van Heerden et al., is available at http://www.dotd .louisiana.gov/administration/teamlouisiana/. The quotes from Kerry Emanuel, Robert Bea, Mark Davis, Pier Vellinga, John Barry, Sandy Rosenthal, Hassan Mashriqui, Beverly Wright, Ivor van Heerden, Don Riley, Robert Twilley, and Oliver Houck are from author's interviews. The information on the Mississippi River Gulf Outlet lawsuit against the Army Corps of Engineers is drawn from the *New York Times,* April 20, 2009.

Michael Grunwald's explanation of the role of new housing in stimulating Florida's economy was offered in a cover story for *Time,* July 10, 2008. Jeb Bush's interview in *Esquire* appeared in the August 2009 issue. The insurance industry's payouts for the 2004 and 2005 hurricane seasons are described by University of Minnesota business professor and insurance expert Andrew Whitman at www.csom.umn.edu/page5555.aspx. Much of the other background information on Florida's insurance crisis was provided by Bill Newton in an author's interview, as well as by contemporaneous news coverage. Information on the Citizens Property Insurance Corporation is available at https://www .citizensfla.com/. For information on the federal flood insurance program, see the program's website, http://www.fema.gov/business/nfip/, as well as a report by the Congressional Research Service that outlines criticisms of the program: http://ncseonline.org/ nle/crsreports/briefingbooks/oceans/p.cfm. Paul Farmer was first quoted in *Time,* August 20, 2006.

The OECD report that likens Shanghai's flood defenses to London's, *Ranking of the World's Cities Most Exposed to Coastal Flooding Today and in the Future,* was published in 2007 and is available at http://www.oecd.org/dataoecd/16/10/39721444.pdf.

The data on Hurricane Gustav, as well as the "lucky" quote from Mark Davis, came from an article in *Time,* September 4, 2008.

Chapter 7: *In Vino Veritas:* The Business of Climate Adaptation

Wayne Leonard's remarks at the Obama White House about climate change are available at http://www.entergy.com/about_entergy/speeches.aspx. Biographical information on Chris West was supplied in an author's interview. Greg Jones's paper, "Extreme Heat Reduces and Shifts United States Premium Wine Production in the 21st Century," is available at http://www.pnas.org/content/103/30/11217.full. More information on wine and climate change, especially concerning Spain, can be found at the Wine Academy's website: http://www.thewineacademy.com/. The information on Lageder's October 2005 wine and climate change conference was drawn from my interview with Lageder and from his company's written report on the conference. The information on the Iceman is drawn from repeated visits to the South Tyrol Museum of Archaeology in Bolzano, Italy, which houses his remains.

The voluminous work of UKCIP in regard to helping businesses adapt to climate change can be accessed at http://www.ukcip.org.uk. The many activities of the Investor Network on Climate Risk are described at http://www.incr.com. The Carbon Disclosure Project's work is at https://www.cdproject.net/en-US/Pages/HomePage.aspx. The

Managing the Unavoidable report is available at http://www.henderson.com/content/sri/ publications/reports/managing_the_unavoidable_final_2009.pdf.

The passage on the insurance industry, especially regarding its activities in the 1990s, is based partly on reporting I did for *Earth Odyssey*. Florida homeowners' loss of insurance and the climate-related actions of the Association of British Insurers and of the other insurance companies mentioned in the text are described in the reports Evan Mills writes every year for the Investor Network on Climate Risk; see, for example, http://www.ceres.org/Document.Doc?id=417.

Chapter 8: How Will We Feed Ourselves?

The description of Mrs. Obama's organic garden is based on contemporaneous news accounts, especially in the *Washington Post*. The description of the Green Revolution's achievements is based in part on *The Doubly Green Revolution: Food for All in the 21st Century* by Gordon Conway (Ithaca: Cornell University Press, 1999). The number of underwater dead zones is documented in "Spreading Dead Zones and Consequences for Marine Ecosystems," by Robert J. Diaz and Rutger Rosenberg, *Science* 321 (August 15, 2008), pp. 926–29. Agriculture's responsibility for 31 percent of global greenhouse gas emissions is documented in "Farming and Land Use to Cool the Earth" by Sara Scherr and Sajal Sthapit, in *State of the World 2009: Into a Warming World* by the Worldwatch Institute (New York: W. W. Norton, 2009), as is meat production's responsibility for 18 percent of emissions. The IMF and World Bank estimates on the role of biofuels in sharpening hunger are reported on pages xvii–xviii of Tristram Stuart's book *Waste* (New York: Norton, 2009). The past effects and projected future impacts of climate change on agriculture in the U.S. Midwest, including that the searing heat of 1988 will become the norm in years ahead, are described in a range of sources: the U.S. government's summary of climate impacts, at http://www.climate.gov/#dataServices/climate AndYou/agriculture; a Union of Concerned Scientists multivolume study, *Confronting Climate Change in the U.S. Midwest*, available at http://www.ucsusa.org/global_warming/ science_and_impacts/impacts/climate-change-midwest.html; studies by the U.S. Department of Agriculture, including http://www.globalchange.gov; and author's interviews with Don Wuebbles, Katharine Hayhoe, and Jerry Hatfield.

The passage on re-greening Africa is based overwhelmingly on my own reporting—the many interviews I did with Chris Reij were particularly helpful—but I first heard about the phenomenon in an article that remains a foundation source of both information and scholarly documentation: "Turning Back the Desert: How Farmers Have Transformed Niger's Landscapes and Livelihoods," in *World Resources 2008: Roots of Resilience: Growing the Wealth of the Poor*, edited by Phillip Angell (Washington, DC: World Resources Institute, 2008). Part of my goal in visiting neighboring Burkina Faso and Mali was to investigate whether the practices of re-greening in Niger had spread, as some advocates had claimed, beyond Niger's borders; they had. The reference to Timbuktu as the hottest city on the planet was based on that day's temperature readings in a copy of *USA Today* that, amazingly, I came across a few days later in Burkina Faso. The study led by Marshall Burke, "Shifts in African Crop Climates by 2050, and the Implications for Crop Improvement and Genetic Resources Conservation," was published in *Global Environmental Change* 19, no. 3 (August 2009): 317–25 and is available at http://www.sciencedirect.com/science?_ob=ArticleURL&_udi=B6VFV-4WFGRNC-1&_user=10&_coverDate=08%2F31%2F2009&_rdoc=1&_fmt=high&_orig=search&_sort=d&_docanchor=&view=c&_searchStrId=1241340344&_rerunOrigin=google&_acct=C000050221&_version=1&_urlVersion=0&_userid=10&md5=6a15e1b20ec69a575bf7f931cc1e3b7c. The story about Malawi appeared in the *New York Times* on April 3, 2007. Some of the activities of Sahel Eco are described on the group's website, http://www.sahel.org.uk/mali.html; a staff member accompanied us on parts of our tour of rural Mali. The Malian government's legalization of tree ownership by farmers was described in author's interviews with numerous Malian NGO activists and Reij. Tony Rinaudo was an early champion of re-greening who graciously agreed to two author's interviews and sent me dozens of his background papers on the subject. The satellite photos supervised by G. Gray Tappan can be viewed at http://lca.usgs.gov. Regarding Chris Reij's estimates of 200 million trees and 12.5 million acres of land, recall that much of the scholarly documentation for the re-greening exercise can be found in the references to the WRI report previously cited, "Turning Back the Desert." Organizers of the Millennium Villages describe their work at http://www.unmillenniumproject.org/mv/index.htm. Norman Borlaug's quote was reported by Gregg Easterbrook in the January 1997 issue of the *Atlantic*. The study by UNEP and UNCTAD, *Organic Agriculture and Food Security in Africa,* is available at http://www.unctad.org/en/docs/ditcted200715_en.pdf.

If only because it reframes the critical question of who was the greatest mass murderer of the twentieth century, *Hungry Ghosts: Inside Mao's Secret Famine* by Jasper Becker (New York: Holt, 1998) ranks among the most important nonfiction books of recent years. The full report by Lin Erda et al. and Greenpeace China is available only in Chinese—I had it translated—but a summary in English is available at http://www.greenpeace.org/china/en/press/reports/climate-food-report-summary. Vaclav Smil made his comments in an author's interview, as did Erda, Ma Shiming, and most of the sources quoted in the remainder of the China passage. Richard Evans's quote is from "China Bets on Massive Water Transfers to Solve Crisis" in *World Rivers Review* 22, no. 4 (December 2007), which is available at http://www.internationalrivers.org/en/node/2397. Premier Wen Jiabao's comments at the 2009 Party Congress on March 5, 2009, were reported by the Xinhua news agency; see http://news.xinhuanet.com/english/2009-03/05/content_10946914.htm. The full citation for Peter Gleick's water study, which is the source of much of the background material on China's water situation and the South-to-North Water Transfer Project, is *The World's Water 2008–2009: The Biennial Report on Freshwater Resources* by Peter Gleick (Washington, DC: Island Press, 2008).

Monsanto's advertisement can be viewed on the company's website: http://www.monsanto.com/pdf/sustainability/advertisement_now_what.pdf. The proportion of profits related to transgenic seeds and the amount of transgenic corn and soybeans in the United States were reported in "Harvest of Fear" by the excellent investigative journalists Donald Barlett and James B. Steele in *Vanity Fair,* May 2008. The WEMA project is described at http://www.monsanto.com/monsanto_today/2009/pledge_wema.asp. The UCS study, *Failure to Yield: Evaluating the Performance of Genetically Engineered Crops,* is at http://www.ucsusa.org/food_and_agriculture/science_and_impacts/science/failure-to-yield.html. Robert Watson's testimony is at http://www.agassessment.org/docs/WatsonTestimony514082.pdf. Sara Scherr's comments about the sequestration capacity of agriculture are drawn from both an author's interview and her article in *State of the World 2009,* cited previously. The Rodale Institute's calculation was cited in the latter. Johannes Lehmann is the leading scholar studying biochar; he and his colleagues explain the underlying science and potential applications in a book that is not easy to read but repays the effort: *Biochar for Environmental Management: Science and Technology,* edited by Johannes Lehmann and Stephen Joseph (London: Earthscan, 2009). George Monbiot's initial criticism appeared in the *Guardian* on March 24, 2009; his (grudging) admission of partial error appeared only on his blog at http://www.guardian.co.uk/environment/georgemonbiot/2009/mar/27/biochar-monbiot-global-warming. The flourishing of urban gardens in Detroit was described in "Detroit Arcadia: Exploring the Post-American Landscape" by Rebecca Solnit, *Harper's,* July 2007. The history and achievements of Victory Gardens were described by Fred Kirschenmann in an author's interview.

Chapter 9: While the Rich Avert Their Eyes

The vulnerability of the ports of Oakland and Long Beach (and much else in California) is documented in the Pacific Institute study *The Impacts of Sea-Level Rise on the California Coast* (see http://www.pacinst.org/reports/sea_level_rise/report.pdf) and in the state government's *2009 California Climate Adaptation Strategy* report, available at http://www.energy.ca.gov/2009publications/CNRA-1000-2009-027/CNRA-1000-2009-027-F.PDF. The relative cheapness of ground-floor rents in Dhaka was mentioned to me by numerous local people. The flooding effects of the 2004 rains were described by Basir Miswas, a marine scientist and activist with the NGO Bangladesh Paribesh Andolon. The study by the Bangladesh Centre for Advanced Studies, *Cyclone '91: An Environmental and Perceptional Study,* is available at http://www.bcas.net/Publication/Pub_Index.html; see also the sequel from BCAS: *Cyclone 1991 Revisited.* Atiq Rahman's follow-up remark was delivered at the conference on Community-Based Adaptation BCAS organized in Dhaka in February 2007, which I attended. The post-1991 adaptation activities are described in the government's *Bangladesh 2008 Climate Change Strategy and Action Plan,* available at http://www. moef.gov.bd/moef.pdf/. The relative size of California's economy is based on International Monetary Fund data. The prevalence of poverty and child labor in Bangladesh is described in the government's adaptation plan and in statements and reports by various NGOs, including Oxfam and the Bangladesh Institute of Labour Studies. The BBC report quoting thirteen-year-old Mijan was broadcast on May 29, 2009.

The description of Sacramento's levees is based on author's interviews with Stein

Buer and Jeff Mount, as well as my own eyewitness observations. Mount also contributed to an essential source of background information: *Envisioning Futures for the Sacramento San Joaquin Delta* (Sacramento: Public Policy Institute, 2007). The vulnerability of Folsom Dam was mentioned in my interviews with Mount, Peter Gleick of the Pacific Institute, and Robert Wilkinson of the University of California, Santa Barbara, and in the essential book on California's water challenges, *Cadillac Desert: The American West and Its Disappearing Water* by Marc Reisner (New York: Viking, 1986). The scheduled upgrade of Folsom Dam was described by Stein Buer in an author's interview. FEMA's strengthening of levee standards was noted by Stein Buer, Jeff Mount, and others in author's interviews, as was the capacity of the spillway outside of Sacramento. The removal of Mount and others from the California State Reclamation Board, and Governor Arnold Schwarzenegger's office's denial, was reported by the *New York Times* on April 18, 2006. The governor's 2007 flood control plan is described at http://gov.ca.gov/index.php?/press-release/7661/. The characteristics of California's water supply system are described in the state's climate change adaptation plan, cited previously. The vulnerability of the delta levees is well known to California officials and was explained to me in interviews with Mount, Maurice Roos, and Robert Wilkinson. The delay in levee repairs in 2009 was reported in the *Sacramento Bee* on July 3, 2009. The cost estimate for building a Peripheral Canal is documented in a study by the Public Policy Institute of California; see http://www.ppic.org/content/pubs/report/R_708EHR.pdf. The Department of Water Resources' estimates of the role of snowpack water and its diminishment by 2100 were conveyed by John Andrews in an author's interview; see also the DWR's report *Managing an Uncertain Future: Climate Change Adaptation Strategies for California's Water* at http://www.water.ca.gov/climatechange/docs/ClimateChangeWhitePaper .pdf. The percentages for where California gets its water came in an author's interview with Robert Wilkinson. Melinda Burns's article about the 20 by 2020 Plan was published in the online magazine *Miller-McCune* on July 29, 2009. The Pacific Institute study, *More with Less: Agricultural Water Conservation and Efficiency in California,* is available at http://www.pacinst.org/reports/more_with_less_delta/more_with_less.pdf. The history and capacity of the Hetch Hetchy dam are described by the Bay Area Water Supply and Conservation Agency at http://bawsca.org/water-supply/hetch-hetchy-water-system/.

Henry Kissinger's "basket case" dismissal of Bangladesh was published in *Time* magazine, January 17, 1972. The information on the killings and other actions undertaken by the West Pakistani army were described in diplomatic cables sent by U.S. embassy staff in Pakistan at the time who dissented from Kissinger's policy; see http:// en.wikipedia.org/wiki/Archer_Blood. The UNDP's ranking of Bangladesh's vulnerability to disasters is found in its report *Reducing Disaster Risk: A Challenge for Development,* available at http://www.undp.org/cpr/whats_new/rdr_english.pdf. The impacts of sea level rise are described in the country's 2008 climate adaptation plan, cited previously. I attended the 2007 conference in Dhaka where Ian Burton made his comments; see also the book *Climate Change and Adaptation* by Ian Burton et al. (London: Earthscan, 2007). The monograph on urban adaptation written by Saleemul Huq and colleagues, *Adapting to Climate Change in Urban Areas,* is available at http://

www.iied.org/pubs/display.php?o=10549IIED. The Bangladeshi government's adaptation activities and spending are described in the 2008 adaptation plan, cited previously. The efforts of the Bangladesh Rice Research Institute were reported by the BBC on December 16, 2009. Todd Stern's rejection of "climate debt" was reported by Reuters on December 9, 2009. The estimate of the costs of adaptation was first published in the World Bank's 2006 study *Clean Energy and Development: Towards an Investment Framework,* available at http://siteresources.worldbank.org/DEVCOMMINT/Documentation/20890696/DC2006-0002(E)-CleanEnergy.pdf. See also a 2010 study coordinated by the World Bank that estimated costs of $75 to $100 billion a year: *The Costs to Developing Countries of Adapting to Climate Change: New Methods and Estimates: The Global Report of the Economics of Adaptation to Climate Change Study,* available at http://siteresources.worldbank.org/INTCC/Resources/EACCExecutiveSummaryFinal.pdf. The study led by Martin Parry, *Assessing the Costs of Adaptation to Climate Change: A Critique of the UNFCCC Estimates,* is available at http://www.iied.org/pubs/display.php?o=11501IIED. John Vidal's article appeared in the *Guardian* on February 20, 2009. I witnessed the speech mentioning the $100 billion climate funding for developing countries that Secretary of State Hillary Clinton delivered in Copenhagen. The National Defense University's exercise was reported in the *New York Times* on August 8, 2009. I witnessed Eileen Claussen's "we want to help you" statement at the Podesta war game.

Chapter 10: "This Was a Crime"

Probably the most influential of Schellnhuber's scientific papers on tipping points, "Tipping Elements in the Earth's Climate System," was published in *Proceedings of the National Academy of Sciences* in 2009 and can be accessed (including an interview with his coauthor, Tim Lenton) at http://sciencewatch.com/dr/nhp/2009/09julnhp/09julnhp LentET/. Schellnhuber described his relationship with Merkel, which was confirmed by other German climate experts speaking on background, and her role in pressing G8 leaders on the 2°C limit in a number of author's interviews. The WBGU study, *Solving the Climate Dilemma: The Budget Approach,* is available in English at http://www.wbgu.de/wbgu_sn2009_en.pdf. Be advised, however, that its prose is hard to penetrate; had I not heard Schellnhuber summarize the study in his presentation in Santa Fe, I never would have guessed that the study carried such powerful conclusions. What I write in this book is therefore based on Schellnhuber's presentation, and subsequent author's interviews, more than on the underlying study. The study by James Hansen and colleagues concluding that 350 ppm was the maximum consistent with avoiding catastrophic climate change, *Target Atmospheric CO_2,* is available at http://www.columbia.edu/~jeh1/2008/TargetCO2_20080407.pdf. The Met Office's study, *Four Degrees and Beyond,* is available at http://www.metoffice.gov.uk/climatechange/news/latest/four-degrees.html.

The rates of greenhouse gas emissions growth are documented in the *Solving the Climate Dilemma* study, among many other sources. The actions the George W. Bush administration took to block progress against climate change were reported at the time by myself and many other journalists; one of the books that describes them in detail is *Censoring Science: Inside the Political Attack on Dr. James Hansen and the Surprising Truth About Global Warming* by Mark Bowen (New York: Penguin, 2007). A much more extensive list of Bush administration actions was compiled by the Natural Resources Defense Council; see http://www.nrdc.org/BushRecord/.

The opposition in the 1990s of Democrats (as well as Republicans) to limits on greenhouse gas emissions was mentioned in author's interviews with Everett Ehrlich, Tim Wirth (who served as undersecretary of state in the Clinton administration), and Philip Clapp, a former Senate aide who later headed the NGO the National Environmental Trust. The 95 to 0 Senate vote is commonly but wrongly described as a vote "against" the Kyoto Protocol. In fact, the vote was taken prior to the Kyoto negotiations, though opponents of climate action subsequently succeeded in convincing many that it was an explicit repudiation of Kyoto. Besides Leggett's *The Carbon War,* other key sources on the corporate disinformation campaign in general and the Global Climate Coalition in particular are Ross Gelbspan's *The Heat Is On,* cited previously, and a website maintained by Greenpeace, ExxonSecrets: http://www.exxonsecrets.org/.

It was that website that first alerted me to Frederick W. Seitz's relationship with tobacco companies, including the (at least) $45 million in research money that the R. J. Reynolds Tobacco Company channeled through Seitz's office at Rockefeller University and his personal receipt of at least $585,000 in return, though not until I interviewed Seitz myself was there confirmation from him that these relationships and payments had indeed taken place. I later arranged for Seitz to be interviewed by PBS *Frontline* for a program called "Hot Politics" on which I was a consulting producer, but the editors of the program, which was broadcast in 2007, chose to omit virtually everything Seitz said, along with my own on-camera comments about corporate influence in Washington. The "doubt is our product" tobacco industry memo was reported in *The Cigarette Papers* by Stanton Glantz et al. (Berkeley: University of California Press, 1998), among other sources. The tenth-anniversary edition of Bill McKibben's *The End of Nature* was published in the United States by Anchor in 1999. The Associated Press investigation of hacked e-mails, written by reporters Seth Borenstein, Raphael Satter, and Malcolm Ritter, "Science Not Faked, but Not Pretty," ran on December 12, 2009.

NASA's findings and the compatibility between the established science of climate change and the conditions of the winter of 2009–10 are described by Joseph Romm in his *Climate Progress* post of April 12, 2010. A list of all national science academies to endorse the science of climate change is available at http://en.wikipedia.org/wiki/Scientific_opinion_on_climate_change. The cartoon by Toles was published in February 2010 and is available at http://climateprogress.org/2010/02/22/toles-on-scientific-uncertainty/. Schellnhuber, Hansen, and Gelbspan all used the word *crime* in our interviews.

BP's $8 billion a year of investments in renewable energy was cited by a senior BP official whom, under the rules of our conversation, I cannot identify by name. He also admitted that BP's investments in traditional fossil fuels dwarfed that amount. For additional information on BP, see the Source Watch report at http://www.sourcewatch.org/index.php?title=BP. Shell's cessation of investments in renewable energy sources was reported in *Grist*, March 18, 2009. The company's ads can be viewed on its website: http://www.shell.com/. Greenpeace describes the role ExxonMobil, the American Petroleum Institute, and the Heritage Foundation played in making misleading claims about climate legislation by reprinting an internal API memo at http://www.desmogblog .com/sites/beta.desmogblog.com/files/GP%20API%20letter%20August%202009-1.pdf/. The grassroots campaign against new coal-fired power plants and the coinciding actions on Wall Street and among leading American politicians is best described by Lester Brown of the Earth Policy Institute at http://www.earthpolicy.org/index.php?/book_ bytes/2010/pb4ch10_ss3/.

The Center for American Progress's analysis of the stimulus package is available at http://www.americanprogress.org/issues/2009/02/recovery_plan_captures.html/. The endorsement of programs titled or similar to a Global Green Deal is documented by Reuters reports on November 6 and December 11 of 2008 and an Agence France-Presse report on February 9, 2009. Gore and Ban's article appeared in the *Financial Times* on February 16, 2009. Much of the material on Amory Lovins is drawn from his article "Climate: Eight Convenient Truths," available at http://www.rmi.org/rmi/Library/ 2009-12_ClimateEightConvenientTruths. George Monbiot's book is *Heat: How to Stop the Planet Burning* (London: Allen Lane, 2006). The New Economics Foundation report is available at http://www.neweconomics.org/publications/growth-isnt-possible/. Wolfgang Sachs and Simon Retallack made their comments at the Ankelohe Conversations, cited in the notes for Chapter 2. The relative carbon footprints of residents of the United States and Burkina Faso were calculated according to figures contained in *Solving the Climate Dilemma: The Budget Approach*, cited previously. Zhou Dadi's energy efficiency studies were described to me in an interview for *Earth Odyssey*. Jiang Kejun's energy efficiency studies were part of the Chinese government's *2050 China's Energy and CO$_2$ Emissions Report*, as described by Reuters in a dispatch published on August 19, 2009. Wolfgang Tiefensee's quote was reported in a special Energy supplement to the *Atlantic Times*, November 2008. The bias of government subsidy programs is documented by a number of sources, including *Estimating US Government Subsidies to Energy Sources, 2002–2008*, by the Environmental Law Institute, available at http://www .elistore.org/Data/products/d19_07.pdf; *Energy Subsidies in the European Union: A Brief Overview*, by the European Environment Agency, available at http://www.eea.europa.eu/ publications/technical_report_2004_1/; and a Reuters dispatch on September 24, 2009, reporting the G20 decision. The Royal Society's report, *Geoengineering the Climate: Science, Governance and Uncertainty*, is available at http://royalsociety.org/geoengineering climate/. The features of cap-and-dividend are explained at http://www.capanddividend .org/, which also links to the *Scientific American* article.

Saudi Arabia's "satisfied" reaction to the Copenhagen climate summit was reported by the BBC on January 4, 2010. Most of the remaining descriptions of what happened at Copenhagen are based on my eyewitness reporting (see the articles and blog postings on http://www.markhertsgaard.com for additional details). The final text of the Copenhagen Accord and other useful information on what was decided at the summit are available at http://unfccc.int/meetings/cop_15/items/5257.php. Milliband's accusations against China appeared in an article he wrote for the *Guardian,* published on December 20, 2009. Mark Lynas published his article in the *Guardian* on December 22, 2009. The accusations were rejected by China; see my article "The Copenhagen Disaccord" in the *Nation,* January 7, 2010. The shortcomings of the climate aid package promised in Copenhagen have been documented by a number of NGOs; see, for example, the World Resource Institute's analysis, *Summary of Climate Finance Pledges Put Forward by Developed Countries,* available at http://www.wri.org/stories/2010/02/summary-climate-finance-pledges-put-forward-developed-countries/. Nicholas Stern's observation about the consequences of a collapse of the climate system was offered in his book *The Global Deal,* cited previously. Joseph Stiglitz explains his cost estimate in *The Three Trillion Dollar War: The True Cost of the Iraq Conflict,* coauthored by Linda J. Bilmes (New York: W. W. Norton, 2008). The $250 billion annual cost of U.S. military bases is estimated in "The Cost of the Global U.S. Military Presence" by Anita Dancs, *Foreign Policy in Focus,* July 2, 2009. McKibben's quote is based on a study published in October 2008 by the National Priorities Project, *The Hidden Costs of Petroleum,* and is found on page 145 of his book *Eaarth,* already cited.

The 650 (and counting) campuses committed to climate neutrality are part of the American College and University Presidents' Climate Commitment; for a complete list of those taking part, see http://www.presidentsclimatecommitment.org/about/commitment/. Lester Brown's observations were made in the article "U.S. Headed for Massive Decline in Carbon Emissions," available at http://www.earth-policy.org/index.php?/plan_b_updates/2009/update83/. Barack Obama made his "change does not happen" statement on January 21, 2008, during a Democratic candidates' debate prior to the South Carolina primary election; a transcript is available at http://www.nytimes.com/2008/01/21/us/politics/21demdebate-transcript.html?pagewanted=all/.

Index